Under RAPs

Under RAPs

Toward Grassroots Ecological Democracy in the Great Lakes Basin

Edited by
John H. Hartig
and
Michael A. Zarull

Ann Arbor
THE UNIVERSITY OF MICHIGAN PRESS

1995   1994   1993   1992   4   3   2   1

Library of Congress Cataloging-in-Publication Data

Under RAPs : toward grassroots ecological democracy in the Great Lakes
   Basin / edited by John H. Hartig and Michael A. Zarull.
      p.   cm.
   Includes bibliographical references and index.
   ISBN 0-472-10258-3 (cloth : alk. paper). — ISBN 0-472-08175-6
(pbk. : alk. paper)
      1. Water quality management—Great Lakes Watershed.   2. Water—
Pollution—Great Lakes Watershed.   I. Hartig, John H., 1952–
II. Zarull, Michael A.
   TD223.3.U53   1992
   363.73'946'0977—dc20   91-41479
                                                                        CIP

Webster's Third International Dictionary
defines *rap* as the legal responsibility
for and the consequences of a criminal act

Foreword

Gro Harlem Brundtland,
Prime Minister of Norway,
Chair, United Nations World Commission
on Environment and Development

For countries having common resources like rivers and lakes, it has always been of interest to establish regimes which could be able to regulate the use of these resources. The United States and Canada have been pioneers in establishing such regimes based on common interest and responsibility.

In its report Our Common Future, the World Commission on Environment and Development underlined the importance of managing our common resources in a way that will satisfy the needs of the present generation without compromising the ability of future generations to satisfy their needs. This is what sustainable development is all about.

Sustainable development is an ambitious concept. It implies that we must transform the way we organize our societies and transcend the stage of just trying to cope with the rapid changes taking place. We must anticipate and prevent, and we must seize control of our future through a new and more active management of global change. We need solutions that can work. Changes depend on whether we are able to address the problems in a comprehensive and systematic manner, and to produce and implement forward-looking decisions.

Democracy and broad participation from all sectors of society are key prerequisites for global change. We need a strong and conscious public opinion to keep democratic pressure on political decision-making alive.

In order to make decisions that will improve the environment, governments depend upon an environmental movement in the population that is strong enough to support even the most difficult decisions. In fact, all political decisions must be supported—or have the potential support of—the majority of the people. Only then can truly effective change come about.

There is a close linkage between ecology and economy. The industrialized countries have been developing for decades without paying for the damage done to the environment. Economic growth has been seen as the leading indicator of development. We cannot continue to measure growth as we have been used to. It is essential that we fully recognize the ecological dimension when searching for new answers and solutions. Growth that degrades the environment is not progress, but deterioration.

We must integrate environmental concerns into all levels of economic planning, performance and accounting. A truly effective strategy for change must be built on a cradle-to-grave approach, from scientific exploration and technological innovation, through the cycles of production, transportation and consumption, to emissions control and waste disposal.

Industry is perhaps the leading instrument of change that affects the environmental resource base of world development. It has been a main cause of air, water and soil pollution, of resource depletion and of dangerous waste material. But industry also possesses the capability to help us find cleaner, safer technologies and enhance the resource base and extend its use.

Sustainable industry requires a transformation of corporate culture which includes the environment as a core value. This can only be achieved through a broad interaction between industry and other sectors of society. Industry must join forces with governments in developing an environment for growth that also fosters growth for the environment, and governments must design framework conditions that help rather than hinder sustainable development.

We need policies that encourage environmentally sound investments and make them profitable. We must make better use of the market to give us a cleaner environment, more quickly and at less cost. Left to itself, the market is a very inadequate instrument for environmental management. In our present economic system, market prices do not reflect the environmental costs of exploitation, production, consumption and waste management. I believe much can be done by combining the effects of standards, emissions limits and new economic instruments.

We experience that the nation state is too small a scene for addressing regional and global environmental degradation. In an age of rapidly growing global interdependence, the traditional nation state is increasingly unable to tackle the challenges of modern civilization alone. To meet the environmental challenges facing us, we need both regional, bilateral and international cooperation.

The Great Lakes represent an important resource to the United States and Canada. Over twenty million residents depend on the Great Lakes for drinking water, they support considerable industrial production, and they are the world's largest inland water transportation network. To foreigners the Great Lakes are known as a beautiful recreation area. Unfortunately, human activities in the area have caused severe damage to the environment.

The RAP program is a result of common concern and shared responsibility for the important resource that the Great Lakes represent. By joining forces from all sectors of society, it has proved possible to identify common goals and to develop common actions to restore the degraded areas. The regimes already existing have proved to be able to meet the new environmental challenges. Information about the RAP program should be of interest far beyond the borders of the United States and Canada, and experiences made from this program should be useful knowledge in our efforts to secure our common future.

Preface

The information presented in this book is based mainly on a series of papers and discussions which constituted a special symposium. This symposium was convened as part of the 34th Conference of the International Association of Great Lakes Research, held at Windsor, Ontario in May 1990.

The development of remedial action plans (RAPs), and, therefore, this book, would not have been possible without the continued dedication of many stakeholders, technical advisers and RAP Coordinators. In addition to the individual chapter authors, we wish to acknowledge the special contributions of Shirley Colthurst, Sally Leppard, Maureen Smyth, Sally Cole-Misch, the staff of the University of Michigan Press, Mike Huggard, Rose Menyes, Rick Coronado, Yvan Gagne and especially Mary Ann Morin.

This work was supported by a grant from the C.S. Mott Foundation and managed through the Citizens Environment Alliance in Windsor, Ontario.

Finally, we would like to thank our families for their patience, enthusiastic support and constant encouragement.

Contents

Foreword   vii
Gro Harlem Brundtland

Preface   xi

Introduction
William K. Reilly   1

Chapter

1. A Great Lakes Mission   5
   John H. Hartig and Michael A. Zarull

2. From Plan to Action: The Green Bay Experience   37
   Victoria A. Harris

3. Hamilton Harbour Remedial Action Planning   59
   G. Keith Rodgers

4. Restoring the Rouge   73
   Roy Schrameck, Margaret Fields, and Margaret Synk

5. A "Two-Track Strategy" for the Buffalo River
   Remedial Action Plan   93
   Barry Boyer and John McMahon

6. Dredging Up the Past: The Challenge of the Ashtabula
   River Remedial Action Plan   121
   Julie A. Letterhos

7. The Quest for Clean Water: The Milwaukee Estuary
   Remedial Action Plan   139
   *Dan Kaemmerer, Audrey O'Brien, Tom Sheffy, and
   Steve Skavroneck*

8. An Overview of the Modeling and Public
   Consultation Processes Used to Develop the Bay of
   Quinte Remedial Action Plan   161
   *Frederick Stride, Murray German, Donald Hurley,
   Scott Millard, Kenneth Minns, Kenneth Nicholls,
   Glenn Owen, Donald Poulton, and Nellie de Geus*

9. Rochester Embayment's Water Quality Management
   Process and Progress, 1887–1990   185
   Margaret E. Peet, Richard Burton, Richard Elliott,
   Philip Steinfeldt, John Davis, Sean Murphy, and
   Andrew Wheatcraft

10. Working toward a Remedial Action Plan for the
    Grand Calumet River and Indiana Harbor
    Ship Canal   211
    Michael O. Holowaty, Mark Reshkin, Michael J.
    Mikulka, and Robert D. Tolpa

11. Remediating Contamination in the Waukegan,
    Illinois, Area of Concern   235
    Philippe Ross, LouAnn Burnett, and Cameron Davis

12. Remediating Polychlorinated Biphenyl
    Contamination in Kalamazoo River/Portage Creek
    Water, Sediment, and Biota   251
    Chris Waggoner and William Creal

13. Keystones for Success   263
    John H. Hartig and Michael A. Zarull

    Glossary   281

    Index   287

Introduction

William K. Reilly, Administrator
U.S. Environmental Protection Agency

Two decades ago, the Great Lakes were severely troubled waters. These magnificent bodies of water were choking on oily wastes, their oxygen depleted by excess nutrients and algae, their fish and wildlife contaminated by pollutants. The Great Lakes were in danger of losing their vitality and their productivity as natural resources.

Fortunately, the United States and Canada, the two nations responsible for the well-being of the Great Lakes, responded to the challenge. This book chronicles the joint efforts of the two countries to restore environmental integrity to the most contaminated parts of the Great Lakes ecosystem. The success of these efforts illustrates the extent to which commitments and investments over the past 20 years are paying off in better environmental quality. But the Great Lakes story also documents unfinished business. It shows how much more remains to be done to bring the Great Lakes back to environmental health.

Environmental cleanup in the Great Lakes basin and in much of North America began in earnest in 1972 with the passage of the U.S. Clean Water Act. That same year, the United States and Canada signed the Great Lakes Water Quality Agreement. The International Joint Commission, created by the Boundary Waters Treaty of 1909, oversees progress in carrying out the Agreement.

The two decades since 1972 saw many additional laws and programs enacted—to curb air pollution, clean up abandoned waste sites, and assure quality drinking water. These efforts have accomplished a great deal; since 1982 we have reduced toxic loadings from hazardous waste sites into the Niagara River by 80 percent, and by 1995 we will have achieved 90 percent removal. Ninety-five percent of United States citizens living in the Great Lakes region are served by waste treatment facilities that remove 85 percent of biological oxygen demand, and the remaining jurisdictions are under court order to get there. Nowhere else on earth has a body of water the size of the Great Lakes shown such remarkable recovery.

Ironically, solving the problems of excess nutrients, oxygen depletion and oily wastes has revealed even more insidious problems caused by chronic exposure to low levels of persistent toxic substances. We also have become increasingly aware of problems caused by the loss of aquatic and terrestrial habitat.

Twenty years ago the causes of environmental decline were relatively obvious. The primary sources of pollution were municipal and industrial discharges. The solution was also fairly straightforward; control what was coming out of the pipes. The remaining problems, on the other hand, are far more subtle and complex. Chronic effects on human and ecological health are less obvious and often are separated from their sources in both distance and time. Sources of toxic substances are diverse and diffuse, for the consequences of human activity are conveyed to the Great Lakes not only by pipes, but by land runoff, groundwater, airborne deposition and other pathways. Our ability to detect pollutants and to identify their effects on living systems is rapidly improving; whereas a generation ago we could measure chemical concentrations in the parts per billion, today we can test for parts per quadrillion for many substances. We know that some toxic chemicals, especially persistent toxic substances that build up in living organisms, can cause damage even in small amounts. Clearly, the discharge of some of these substances into the ecosystem should cease entirely—much as we banned DDT, dieldrin, toxaphene, and mirex—or be significantly reduced.

Traditional laws and programs that address single source categories or single media such as air or water cannot solve the problems by themselves. We need to protect a complex, dynamic living system. To protect it effectively, we must focus on tangible environmental results measured in terms of human and ecological health throughout the entire ecosystem. And we must make a major commitment to prevent pollution before damage is done.

This book describes one such integrated, geographical approach to solving complex problems in "hot spot" areas in the Great Lakes basin. Forty-three of these "hot spots," or Areas of Concern, have been identified—26 in United States waters, 12 in Canadian waters, and five in waters shared by both countries. Toxicity problems showed up in the Great Lakes earlier than in other regions because most of the pollutants entering the lakes stay in them. The waters of Lake Michigan, for example, "turn over" only once every 100 years; in Lake Superior, a complete flushing takes place only every 200 years. The most polluted parts of the lakes are at the mouths of the rivers. They provide excellent locations for harbors, industry and urban development. They are also places where upstream pollutants are deposited as river flows diminish when they enter the lakes.

In these urban industrial areas, contaminated sediments have been accumulating for decades. Contaminated sediments are found world-wide, of course, but because of the sensitivity of our "sweetwater seas," these problems were recognized in the Great Lakes earlier than elsewhere. This had also led to early development of strategies and methods to address the problems, including the key concepts of Areas of Concern and Remedial Action Plans.

In this program, the Great Lakes Water Quality Board of the International Joint Commission defines geographic Areas of Concern where human or ecosystem health is impaired, as measured in terms of "use impairments." Remedial Action Plans identify the steps needed to restore the ecosystem's stability. By 1985, well before the Areas of Concern and Remedial Action Plan provisions were officially added to the Agreement, state and provincial environmental agencies had embraced these concepts. They recognized that comprehensive plans would be needed to address the problems. The early efforts in developing plans by the various jurisdictions were experiments, updated in the political, social, and scientific context of the Great Lakes community.

A prime function of the Remedial Action Plan process is to enable communities to assemble all the programs, resources, and authorities into a complete picture of the area's environmental problems and their ability to respond. Any information gaps, resource needs, or programmatic deficiencies soon become evident. Planning is only the first step; it must be followed by sound, clearheaded analysis of what is doable, both technically and economically. For example, one possible option, managing contaminated sediment in the Great Lakes, is an immense, costly technological challenge; it will require us to push the state of the art in innovative technology. We must look to science—the best science we can muster—to tell us how we might best proceed and how much we reasonably can expect to accomplish.

Another important ingredient to the success of the Remedial Action Plan process, perhaps the most essential, is stakeholder involvement. It creates a formidable constituency of well-informed people who understand the problems and the steps needed to solve them. Involving the regulated community and the public promotes understanding. Assigning responsibility for the actions needed to reach ecosystem goals and the Parties responsible for taking those actions helps ensure accountability.

The "Area of Concern—Remedial Action Plan" process has implications that reach far beyond the Great Lakes basin. This process is providing an example for the United States, Canada and the world. It provides planning that is geographically focused, ecosystem based, and cuts across environmental media. The approach offers a model of how to make better use of existing programs while determining the other actions needed to restore human and ecosystem health. By involving stakeholders, it builds support and creates accountability. It provides a model for successful problem solving and a practical way to attain the goals of the U.S. Clean Water Act and the Great Lakes Water Quality Agreement: to "restore and maintain the chemical, physical and biological integrity of the Nation's waters" and "the chemical, physical and biological integrity of the waters of the Great Lakes Basin Ecosystem."

Along with planning and analysis, U.S. Environmental Protection Agency (EPA) is actively pursuing environmental improvement in the Great Lakes on two other major fronts: enforcement and education. We have brought a number of major enforcement actions against cities and industries in the Great Lakes region that have violated the laws against releasing pollutants into the lakes. In 1991, EPA's Chicago/ Great Lakes regional office assessed more penalties under the Clean Water Act than the entire agency had collected since 1972 under this law. Vigorous and consistent enforcement is the foundation of EPA's integrity as a regulatory agency, in the Great Lakes as elsewhere; polluters must and will pay. To restore and protect the lakes, tough enforcement, strict regulation of pollutants and pollution sources, and ambitious cleanup all will be necessary.

And we will also need education—to instill a new environmental ethic in our citizens, and to develop new environmental literacy and skills in our next generation of scientists, engineers, mathematicians, economists, and other trained specialists. As these technically proficient people become involved in environmental work, more breakthroughs in cleanup and prevention technologies will occur. And more and more companies will see the value of pollution prevention—the economic as well as the environmental wisdom of reducing or eliminating waste at the source.

Realizing our ambitious goals for the Great Lakes will require the best efforts of governments and citizens alike on both sides of the border. It's well worth it. The potential payoff is enormous, not just for the Great Lakes, but in fashioning a model for how we can move forward, from planning to action, to protect and restore all of the nation's great water bodies to full health and productivity. It's one of the most important jobs we have. Let's get on with it.

Chapter 1

A Great Lakes Mission

John H. Hartig and Michael A. Zarull

"An essential element in restoring the integrity of the Great Lakes is the understanding, active support and involvement of the public. Nowhere is this more applicable—or potentially effective—than in remedial action plans. Indeed, the great potential of the remedial action plan process, already demonstrated in several Areas of Concern around the Great Lakes, lies in drawing together the various interests and expertise needed to identify common goals and then the means to resolve local environmental problems in a comprehensive, systematic manner. To be effective, remedial action plans will need to incorporate an understanding of the societal factors that have generated the problem or could benefit from its remediation, as well as preventive measures against further degradation and adequate, long-term funding. Most importantly, while they may require legal, technical and economic leadership from governments, remedial action plans can only be effective if they are reinforced by a broad community commitment to sustainable attitudes and lifestyles."

E. Davie Fulton
Canadian Chairman
International Joint Commission

Introduction

When the last of the great glaciers retreated some 12,000 years ago, they left behind a necklace of five bright blue jewels strung along the east-central portion of North America. This necklace of "sweetwater seas" became known as the Great Lakes because of their sheer size and volume of freshwater; these magnificent lakes cover approximately 246,000 km² (95,000 square miles) and hold almost one-fifth of the total freshwater on the earth's surface (22.71 trillion cubic meters; 6,000 trillion gallons).

The Great Lakes region has long been considered the heartland of North America because of its resources and the role the lakes have played in the history and development of the United States and Canada. For native cultures and early European explorers, the lakes and their tributaries provided basic sustenance and avenues to discover new lands and abundant resources. Today, approximately 30 percent of Canadians (7.5 million people) and 12 percent of United States' population (30 million people) live in the Great Lakes drainage basin. Twenty-four million residents depend on the Great Lakes for drinking water. Because of the availability of abundant, inexpensive water and accessible, efficient transportation, the Great Lakes region continues to support considerable industrial productivity (e.g. 50 percent of United States steel production and 62 percent of Canadian steel production) and provides both countries with the world's largest inland water transportation network.

The Great Lakes represent not only a shared resource between Canada and the United States, but also a shared responsibility for their stewardship. The availability and quality of this resource is at the foundation of the basin's technologically advanced society, and its protection is fundamental to assuring the region's future economic viability. It should come as no surprise that some of the world's largest centers of commerce, industry and urban development are found in the Great Lakes basin; unfortunately, pollution has accompanied this development.

Since early settlement the vastness of the Great Lakes has generated the false impression that they have an infinite capacity to absorb human abuse and waste (Thomas and Hartig, 1989). In fact, human mismanagement of the lakes has produced measurable deleterious effects on water quality and the organisms that live there, including humans, and has shown the fragility of the ecosystem.

Early European explorers found native cultures living in harmony with their ecosystem. In the 1800s, as forests were cleared by Europeans for agricultural land, increased soil loss brought siltation and loss of suitable spawning and nursery ground habitat for fish. Removal of vegetative cover also raised stream temperatures, which in turn dramatically reduced populations of fish that required cool water temperatures.

As settlements developed and prospered along the shoreline, the discharge of human waste resulted in visible contamination of the lakes, as well as startling and sometimes even fatal human contamination. By the early 1900s several epidemics of typhoid fever and cholera resulted from drinking polluted water supplies. Later, oil pollution from industrial growth occurred in the 1940s–1960s, and uncontrolled phosphorus enrichment brought accelerated eutrophication during the 1960s. Today, the effects of hundreds of toxic substances that entered (and many continue to enter) the Great Lakes via industrial and municipal discharges, land runoff, and atmospheric deposition are evident. This toxic contamination has continued to the point that the Great Lakes now recontaminate themselves from their own bottom sediments (Hartig and Thomas, 1988).

Forty-three "pollution hotspots" or Areas of Concern have been identified in the Great Lakes basin where the general or specific water quality objectives of the Canada-United States Great Lakes Water Quality Agreement are not met. In these areas, the pollution has impaired or is likely to impair the beneficial uses of the water body or the area's ability to support aquatic life (Figure 1). Impairment of beneficial use means that the chemical, physical, or biological integrity of the Great Lakes ecosystem has changed sufficiently to cause any or all of the following:

- restrictions on fish and wildlife consumption
- tainting of fish and wildlife flavor
- degradation of fish and wildlife populations
- fish tumors or other deformities
- bird or animal deformities or reproductive problems
- degradation of benthos
- restrictions on dredging activities
- eutrophication or undesirable algae
- restrictions on drinking water consumption, or taste and odor problems
- beach closings
- degradation of aesthetics
- added costs to agriculture or industry
- degradation of phytoplankton and zooplankton populations
- loss of fish and wildlife habitat (Canada and the United States, 1987).

Areas of Concern include rivers (Cuyahoga River, Ohio), harbors (Hamilton Harbour, Ontario), connecting channels (Niagara River, New York and Ontario), and large embayments (Saginaw Bay, Michigan).

As a result of a 1985 recommendation of the International Joint

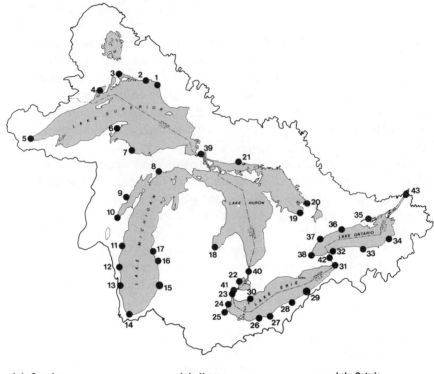

Lake Superior

1 Peninsula Harbour
2 Jackfish Bay
3 Nipigon Bay
4 Thunder Bay
5 St. Louis Bay / River
6 Torch Lake
7 Deer Lake -
   Carp Creek / River

Lake Michigan

 8 Manistique River
 9 Menominee River
10 Fox River / Southern Green Bay
11 Sheboygan River
12 Milwaukee Estuary
13 Waukegan Harbor
14 Grand Calumet River /
    Indiana Harbor Canal
15 Kalamazoo River
16 Muskegon Lake
17 White Lake

Lake Huron

18 Saginaw River / Saginaw Bay
19 Collingwood Harbour
20 Severn Sound
21 Spanish River Mouth

Lake Erie

22 Clinton River
23 Rouge River
24 River Raisin
25 Maumee River
26 Black River
27 Cuyahoga River
28 Ashtabula River
29 Presque Isle Bay
30 Wheatley Harbour

Lake Ontario

31 Buffalo River
32 Eighteen Mile Creek
33 Rochester Embayment
34 Oswego River
35 Bay of Quinte
36 Port Hope
37 Metro Toronto
38 Hamilton Harbour

Connecting Channels

39 St. Marys River
40 St. Clair River
41 Detroit River
42 Niagara River
43 St. Lawrence River
    (Cornwall / Massena)

Fig. 1. Forty-three Areas of Concern identified in the Great Lakes Basin

Commission's (IJC) Great Lakes Water Quality Board, the eight Great Lakes states and the Province of Ontario committed themselves to developing and implementing a remedial action plan (RAP) to restore beneficial uses in each Area of Concern within their political boundaries (IJC 1985). Each RAP must identify when specific remedial actions will be taken to resolve the problems, as well as designating who is responsible for implementing those actions.

The development and implementation of RAPs represents a challenging departure from most historical pollution control efforts, where separate programs for regulation of municipal and industrial discharges, urban runoff, agricultural runoff and dredging for navigational purposes were implemented without considering overlapping agency responsibilities, cumulative impacts, or the interrelatedness of ecosystem components (Hartig and Thomas, 1988). This new process also calls on the talents available from a diverse array of programs that extend far beyond those traditionally associated with water pollution control, including the involvement of local communities and various government agencies at all levels. These programs include—but are not limited to—municipal and industrial wastewater treatment, hazardous waste management, nonpoint source pollution control, groundwater, fisheries and wildlife management, dredging and harbor maintenance, land use planning, and recreation. Thus, all programs, agencies and communities whose actions affect an Area of Concern must work together on common goals and objectives to develop and implement a RAP.

Remedial Action Plans and the International Joint Commission

Canada and the United States have a long history of cooperatively resolving issues affecting their shared environment. A formal process for this cooperation began with the signing of the 1909 Boundary Waters Treaty, which established the International Joint Commission (IJC). This independent body, composed equally of United States and Canadian appointees, provides a quasi-judicial and investigative mechanism to cooperatively resolve problems (including water and air pollution, fluctuating lake levels and other issues) along the two countries' common border.

Waterborne disease epidemics was the first water pollution issue addressed by the IJC in the early 1900s. The IJC identified pollution sources and recommended specific remedial actions, including water purification and treatment, to control the pollution. As a result of the widespread recognition of accelerated eutrophication of the Great Lakes (e.g. Lake Erie) and various IJC studies identifying the pollution problems, the first Great Lakes Water Quality Agreement was signed between Canada and the United States in 1972. The 1972 Agreement gave the IJC specific responsibilities and functions (i.e. to review and assess progress under the Agreement) and provided the focus for a coordinated approach to control eutrophication by reducing phosphorus inputs.

As scientists and society learned more about the Great Lakes, the Agreement was revised and expanded in 1978. This new Agreement recognized the need to understand and effectively manage the discharge of toxic substances into the Great Lakes basin. The United States and Canada, through the 1978 Agreement, also formally adopted an ecosystem approach which attempts to account for the interrelationships among water, air, land and all living things, including humans. This approach provides a more holistic planning, research, and management framework.

In 1987, the Great Lakes Water Quality Agreement was again revised to include atmospheric deposition of toxic substances, contaminated sediments, groundwater, nonpoint source pollution, RAPs, and lakewide management plans. The RAP process was formalized and requires RAPs to be submitted to the IJC for review and comment at three stages: problem definition; selection of remedial actions; and confirmation of use restoration. The RAP program explicitly became part of the Great Lakes Water Quality Agreement and represents the first opportunity, on a broad and practical scale, to restore degraded areas of the Great Lakes using an ecosystem approach. Colborn et al. (1990) suggest that RAPs represent the major test of the ecosystem approach and are the means through which the ecosystem approach will involve and affect large numbers of individuals and institutions.

Development and Implementation of RAPs

The attitudes and uses of the Great Lakes have evolved from settlement and initial development to its current state where RAPs are being developed and implemented, using an ecosystem approach, to restore degraded areas (Table 1; adapted from Hartig and Hartig, 1990). Public awareness and involvement also have evolved from initial, first hand experiences with pollution (e.g. typhoid fever and cholera epidemics in the early 1900s) to public participation in planning, development and implementation of RAPs. This increasing public awareness and stakeholder involvement has led to more effective overall management of the basin.

Perhaps the best way to describe the use of the ecosystem approach in the RAP process is to compare it to the process of simultaneous engineering of automobiles. In 1985, the Ford Taurus and the Mercury Sable were the first North American cars built on the innovative principle of simultaneous engineering. Simultaneous engineering is a team approach to designing, building and selling cars and involves every department, which ultimately bear responsibility for the product. The fundamental principle is to involve every area of the company simultaneously, rather than one at a time; representatives from design, engineering, manufacturing, marketing, finance and suppliers all work together from start to finish. The process breaks down barriers between disciplines, stifles the tendency for empire building, and fosters communication. The right hand, in short, knows what the left hand is doing and also can offer suggestions to help it to better do its job.

What simultaneous engineering is to automobile manufacturing, the ecosystem approach is to restoring degraded areas of the Great Lakes. The ecosystem approach is integrative and holistic and recognizes the complex interrelationships and interdependence among all parts of the system. Although scientists have used and advocated the ecosystem approach to perform research for years (IJC 1978, Francis et al. 1979), regulatory and management agencies have made little progress in implementing the approach (Christie et al. 1986, Harris et al. 1987). This changed in 1985 when the RAP program was established.

For over 10 years the Great Lakes Water Quality Board of the IJC tracked programs and progress in abating bacterial pollution and controlling cultural eutrophication. The Board was not satisfied, however, with progress in resolving the more significant problem of toxic substances contamination. Thirty-nine of the 43 Areas of Concern have restrictions on human consumption of fish, 18 areas have fish tumors, 36 areas have restrictions on dredging activities due to the presence of toxic substances, and 42 have toxic substances in sediments (IJC 1985, 1987a).

Both the problems and solutions associated with these issues are complex. Therefore, the Great Lakes Water Quality Board recommended that the eight Great Lakes states and the Province of Ontario take the lead in developing a comprehensive remedial action plan (RAP) to restore impaired beneficial uses in each Area of Concern within their political boundaries (IJC 1985). Immediately fol-

TABLE 1.   Evolving Management of Degraded Areas in the Great Lakes

Era	Explanation and/or Examples	Approximate Dates
Settlement and development	The fur trade was the principal industry to bring about settlement and colonization of coastal areas.	1600s and 1700s
Exploitation	Clearing of forest land for agricultural purposes and timber caused increased soil loss, which resulted in siltation and loss of spawning and nursery ground habitat for fishes.	1800s
Reactive management	Typhoid fever and cholera epidemics led municipalities to build purification facilities.	Early 1900s
	Cultural eutrophication of Lake Erie led governments to limit the phosphorus content of household laundry detergents and to limit phosphorus in municipal and industrial effluents. Contamination of the fishery led to the banning of DDT, dieldrin, etc.	1960s and 1970s
Proactive management	Waste reduction/minimization and risk/ hazard assessment to prevent toxic substance problems.	1980s and 1990s
Development and implementation of RAPs using an ecosystem approach	Institutional structure is established, broadly representative of social, economic and environmental interests; agreement is reached on problems and use impairments, goals are identified, and specific actions are taken to achieve these goals; work is to continue in perpetuity to achieve continuity of purpose and ensure accountability.	Present

lowing the Water Quality Board's 1985 recommendation, the eight Great Lakes states and the Province of Ontario committed themselves to developing and implementing RAPs.

Each RAP must address the following:

A. define the environmental problem, including the geographic extent using detailed maps and surveillance information
B. identify impaired beneficial uses
C. describe the causes of the problems and identify all known sources of pollutants
D. identify remedial actions proposed to restore beneficial uses
E. identify a schedule to implement remedial actions
F. identify jurisdictions responsible for implementing and regulating remedial actions
G. describe the process to evaluate remedial program implementation and regulating remedial measures and
H. describe the surveillance and monitoring activities that will be used to track program effectiveness and eventual confirmation that uses have been restored.

The Great Lakes Water Quality Agreement states that RAPs "shall embody a systematic and comprehensive ecosystem approach to restoring and protecting beneficial uses in Areas of Concern" (Canada and the United States, 1987). The ecosystem approach thus must involve and consider all users in policy making and management, including scientists, regulators, industry and citizen representatives, and others. The RAP program has been described as a unique experiment in institutional cooperation and an opportunity to implement the ecosystem approach at a "grass roots" level.

The two-dimensional diagram presented in Figure 2 depicts a traditional view of the responsibilities of different agencies, organizations, and programs which must be involved in the RAP program (IJC 1987a). The jurisdictions responsible for developing RAPs are at the center of the diagram, where the vertical and horizontal axes cross. The vertical axis depicts the range of different responsibilities and interests, from international and national government to public interest groups and concerned citizens. The horizontal axis depicts the diverse program responsibilities within each jurisdiction. The

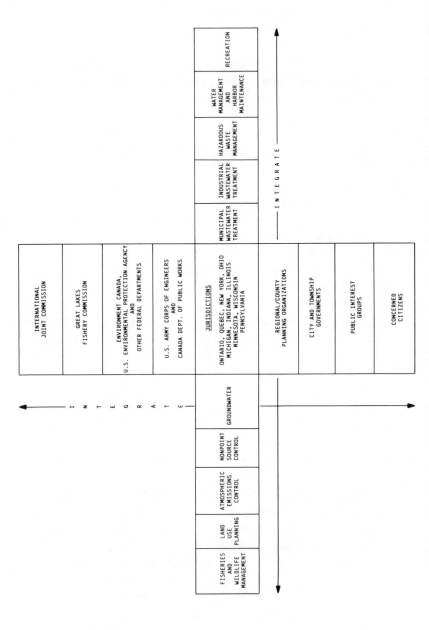

Fig. 2. A two-dimensional schematic diagram which depicts the need to integrate the responsibilities of different agencies, organizations, and programs under the "umbrella" of a Remedial Action Plan (from IJC 1987b)

challenge would be for the eight Great Lakes states and the Province of Ontario to effectively integrate these various responsibilities and interests on the vertical and horizontal axes in RAP development and implementation.

Incorporating the ecosystem approach into the RAP program means viewing the organizations and interests identified in the vertical and horizontal axes as equal members of a team (Figure 3). Thus, the traditional diagram must be rearranged into a circle, with open spaces between compartments (Hartig and Vallentyne, 1989). In this model, each member of the RAP team can communicate with all other members of the team in order to accomplish their common goal to develop and implement a RAP for their Area of Concern. The underlying premise is that multi-media, community based watershed planning is needed to address the interrelationships between ecosystem compartments and the true causes of problems such as toxic substances contamination. Stakeholder ownership of this planning process also should improve the prospects for plan implementation.

The RAP program has been described as "A Great Lakes program whose time has come" because it is changing the way society solves environmental problems. It is fair to say that politicians, government agency personnel, the public and other interest groups are now operating in an open system of resource management—a glass house. While it is difficult for all sides, politicians and government agency personnel are learning to share environmental decisionmaking power. Citizens are learning about the value and limits of science, that all decisions have consequences, and that management is often forced to function with a high degree of uncertainty. This ongoing process of human interaction, in itself, may be a barrier to direct and quick action, but is essential to ensure that all parties understand the issues facing the Area of Concern and are committed to implementing the RAP.

The Great Lakes Water Quality Board intentionally has placed emphasis on action in the RAP process. The purpose of RAPs is to restore impaired beneficial uses through a process which identifies when specific remedial actions will be taken to resolve the problems and which organizations or agencies are responsible for implementation. There is broad agreement throughout the Great Lakes basin that the emphasis of RAPs must be on remedial action

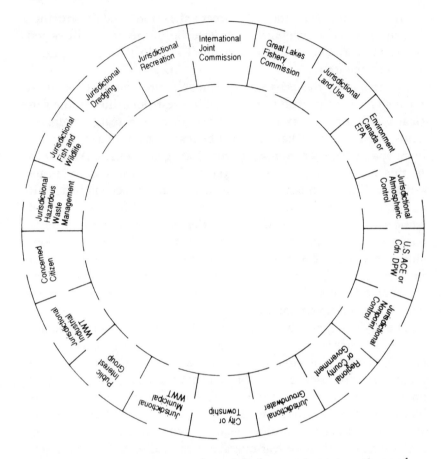

Fig. 3. An ecosystem approach model of promoting interaction and communication between different interests involved in Remedial Action Plan development (from Hartig and Vallentyne 1989)

and that there must be a commitment to complete the job correctly for each Area of Concern.

Cornerstones of the RAP Program

While each RAP has its own unique approach and constraints, there are cornerstones or critical elements for success which are common to all successful RAPs. These are the building blocks of the foundation upon which the rest of the plan is designed and implemented. Presented below are the major cornerstones of the RAP program and selected examples of their effectiveness.

Institutional Structures and Public Participation

One of the cornerstones of the RAP program is the establishment of either a stakeholders' group, basin committee, citizens' committee, or public advisory committee to assist in RAP development and implementation. In 33 of the 43 Areas of Concern (Table 2), these groups broadly represent environmental, economic and social interests, and provide the first opportunity for meaningful public participation in ecosystem management (Hartig et al. 1990a). This participation goes beyond the traditional input received in public hearings and emphasizes listening, learning, discussing, and developing a plan together.

The empowerment of these groups has changed the traditional way of doing business and the level of interaction as a result of the diverse and direct involvement in decisionmaking. Fundamental to the success of the RAP process is trust among all stakeholders. Indeed, broad-based institutional structures and a commitment to a democratic decisionmaking process are essential to a coordinated societal response to environmental problems confronting the Great Lakes Basin Ecosystem (Thomas et al. 1988). These stakeholder groups in Areas of Concern are reaching agreement on problem definition and goals, identifying remedial options, choosing preferred remedial actions, and are working to achieve accountability. The benefits of opening up the RAP process have been achieving greater public outreach, establishing cooperative working relationships, while building public and business support for remediation (Hartig et al. 1991).

The RAP process also has shifted the traditional "top-down" approach to environmental planning, where government agencies in capital cities develop site-specific plans for isolated geographic areas, to a "bottom-up" approach where stakeholders within their Area of Concern are productively involved in their RAP's development and implementation. Experience has shown that stakeholders relate to their Area of Concern because it is where they live and work, thus they have a greater commitment to plan for and ensure

TABLE 2. Institutional Structures Established to Assist in Development of Remedial Action Plans in Areas of Concern in the Great Lakes Basin

Area of Concern (jurisdiction)	Organizational/Institutional Structure(s)	Representation
1. Peninsula Harbor (Ontario)	Public Advisory Committee (PAC)	chamber of commerce, citizens-at-large, industry, interest group, municipal government, recreation
2. Jackfish Bay (Ontario)	Public Advisory Committee (PAC)	business/industry, citizens-at-large, city official, interest group, recreation, tourism, union
3. Nipigon Bay (Ontario)	Public Advisory Committee (PAC)	chamber of commerce, church, citizens-at-large, industry, interest group, municipal government, native group
4. Thunder Bay (Ontario)	Public Advisory Committee (PAC)	business/industry, citizen, city official, conservation authority, environmental group, interest group, municipal government, native group, power generation, tourism
5. St. Louis River (Minnesota/ Wisconsin)	St. Louis River Citizens Advisory Committee (CAC)	academic, business/industry, citizens-at-large, city/county official, environmental groups, native group, power generation, recreation, regional government, state/federal government
6. Torch Lake (Michigan)	Does not currently exist—two public meetings and a public comment period held initially. The desire/need for an institutional structure will be evaluated when the Stage 2 RAP is developed.	

7. Deer Lake—Carp Creek—Carp River (Michigan)	Does not currently exist—two public meetings and a public comment period held initially. The desire/need for an institutional structure will be evaluated when the Stage 2 RAP is developed.	
8. Manistique River (Michigan)	Does not currently exist—two public meetings and a public comment period held initially. The desire/need for an institutional structure will be evaluated when the Stage 2 RAP is developed.	
9. Menominee River (Wisconsin/Michigan)	Menominee River RAP Citizens Advisory Committee (CAC)	academia, business/industry, chamber of commerce, citizens-at-large, city official, county representative, environmental group, federal government, fishing club, interest group
10. Fox River/Southern Green Bay (Wisconsin)	Citizens Advisory Committee (CAC) Implementation Committee (IC)	academia, business/industry, chamber of commerce, citizens-at-large, city/county officials, elected officials, interest group, native group, regional agencies
11. Sheboygan (Wisconsin)	Sheboygan County Water Quality Task Force	citizens-at-large, private industry, sportsmen's groups, state government

(continued on next page)

(Table 2—*continued*)

Area of Concern (jurisdiction)	Organizational/Institutional Structure(s)	Representation
12. Milwaukee Harbor (Wisconsin)	Milwaukee River RAP Citizens Advisory Committee (CAC)	academia, business/industry, chamber of commerce, citizens-at-large, city/county officials, elected officials, interest group, native group, regional agencies
13. Waukegan Harbor (Illinois)	Waukegan Harbor Citizens Advisory Group	city government, industry, environmental groups, health department, municipalities, sport fishery groups, business, recreation, universities
14. Grand Calumet River/ Indiana Harbor Canal (Indiana)	Citizens Advisory for the Remediation of the Environment Committee	academic, business, citizens groups, environmental groups, industry, local government
15. Kalamazoo River (Michigan)	Kalamazoo River Basin Strategy Committee (disbanded in 1989). Future efforts will be coordinated with Superfund. The desire/need for an institutional structure will be evaluated.	business, charter boat owners, county officials, government, industry, property owners, teachers
16. Muskegon Lake (Michigan)	Does not currently exist—two public meetings and a public comment period held initially. The desire/need for an institutional structure will be evaluated when the Stage 2 RAP is developed.	

Area of Concern	Committee	Stakeholder groups
17. White Lake (Michigan)	Does not currenlty exist—two public meetings and a public comment period held initially. The desire/need for an institutional structure will be evaluated when the Stage 2 RAP is developed.	
18. Saginaw River/ Saginaw Bay (Michigan)	Saginaw Basin Natural Resources Steering Committee	academia, agriculture, county officials, environmental groups, municipal officials, regional planning commission
19. Collingwood Harbour (Ontario)	Collingwood Harbour RAP Public Advisory Committee (PAC)	academia, chamber of commerce, citizens-at-large, conservation authority, environmental groups, industry, interest groups, power generation, recreation/tourism, sewage treatment plant
20. Severn Sound Penetang Bay to Sturgeon Bay (Ontario)	Severn Sound Remedial Action Plan Public Advisory Committee (PAC)	citizens-at-large, interest groups, local municipalities, recreation
21. Spanish River Mouth (Ontario)	Public Advisory Committee (PAC)	academia, citizens groups, environmental groups, health, industry, local government, native people, recreation/tourism

(continued on next page)

(Table 2—continued)

Area of Concern (jurisdiction)	Organizational/Institutional Structure(s)	Representation
22. Clinton River (Michigan)	Clinton River RAP Advisory Committee	Clinton River Watershed Council (CWRC) will facilitate meetings including the following groups: academia/business/ industry, chambers of commerce, city officials, environmental groups, federal/state government, health officials, recreation, sewage treatment plants. Other groups involved will be: Lake St. Clair Advisory Committee, East Michigan Environmental Action Council (EMEAC), The Southeastern Michigan Council of Governments (SEMCOG).
23. Rouge River (Michigan)	Executive Steering Committee (ESC) Basin Committee (BC)	Executive Steering Committee: apppointed citizens-at-large, elected official, federal/state government, governor's representatives, House and Senate officials. Basin Committee: business/ industry, environmental group, interest groups.
24. River Raisin (Michigan)	Does not currently exist—two public meetings and a public comment period held initially. An institutional structure will be organized in the near future.	
25. Maumee River (Ohio)	Maumee River Remedial Action Plan Advisory Committee (RA PAC)	academia, business/industry, chamber of commerce, city officials, environmental groups, federal representatives, federal/state government, health officials, interest groups, recreation

Area of Concern	Group	Stakeholders
26. Black River (Ohio)	None—limited resources have prevented formation of a group	
27. Cuyahoga River (Ohio)	Cuyahoga Coordinating Committee (CCC)	citizen, elected official, industry, interest group, municipal
28. Ashtabula River (Ohio)	Ashtabula River RAP Advisory Council	citizen, elected official, industry, interest group, municipal
29. Presque Isle Bay (Pennsylvania)	Erie Harbor Improvement Council	citizen, industry, municipalities, academia, business, conservation groups, city and state government
30. Wheatley Harbour (Ontario)	None—no interest in forming a formal structure. They meet informally and keep informed by newsletters which seek public comment.	
31. Buffalo River (New York)	Buffalo River Citizens' Committee (BRCC)	academia, church, environmental groups, labor groups, local business, local government representatives, public interest groups, sportsman's organization

(continued on next page)

(Table 2—*continued*)

Area of Concern (jurisdiction)	Organizational/Institutional Structure(s)	Representation
32. Eighteen Mile Creek (New York)	RAP has not been initiated. RAP development will begin when Buffalo River and Niagara River RAPs are finished	
33. Rochester Embayment (New York)	Water Quality Management Committee (WQMC)	citizens, economic interests, ex-officio non-voting members, public interests, public officials
34. Oswego River (New York)	Oswego River Remedial Action Plan Citizens' Committee (ORCC)	business/industry, elected officials, environmental groups, governmental agencies, university, public interest groups
35. Bay of Quinte (Ontario)	Public Advisory Committee (PAC)	academia, agriculture, environmental, human health, industry, labor, municipal, naturalist/nature, tourism/recreation
36. Port Hope (Ontario)	Port Hope Harbour Remedial Action Plan Local Advisory Group	concerned citizens, conservation authority, federal siting task force, harbour commission, industry, local government, low-level radioactive waste management office, yacht club
37. Metro Toronto (Ontario)	Public Advisory Committee (PAC)	PAC sectors: agriculture, business/industry, community groups/individuals, environment/conservation, labor, Metro Toronto & Region Conservation Authority, recreation/tourism, Toronto Harbour Commission.

38. Hamilton Harbour (Ontario)	Hamilton Harbour Stakeholders	academia, boat clubs, business/industry, chamber of commerce, city officials, citizens, conservation authority, environmental groups, fed./prov./mun. government, interest groups, union
39. St. Marys River (Ontario/Michigan)	St. Marys River Binational Public Advisory Council (BPAC)	academia, citizens-at-large, environmental groups, fisheries, industry, labor, municipal representatives, native people, public health, recreation/tourism, small business
40. St. Clair River (Ontario/Michigan)	St. Clair River Binational Public Advisory Council (BPAC)	agriculture, business/industry, citizens-at-large, commercial fishery, community groups, conservation & environmental, health, labor, municipal, native people, provincial/state agencies, tourism/recreation
41. Detroit River (Michigan/Ontario)	Detroit River Binational Public Advisory Council (BPAC)	40 members (20 U.S. & 20 Canadian): academia, citizens, conservation/environmental, industry & port authority, labor, municipal, nonpoint sources, recreation
42. Niagara River (Can.) (Ontario)	Public Advisory Committee (PAC) for the Niagara River (Ontario) Remedial Action Plan	academia, agriculture, citizens-at-large, commissions, community group, conservation authority, environmental groups, health, industry, labor, municipal government, power generation, tourism/recreation
42. Niagara River (U.S.) (New York)	Committee of Canadian and U.S. Citizens from Niagara River area	academia, economic interests, government official, labor, private citizen, public interests, researcher

(continued on next page)

(Table 2—continued)

Area of Concern (jurisdiction)	Organizational/Institutional Structure(s)	Representation
43. St. Lawrence River (Can.) (Ontario)	Public Advisory Committee (PAC)	academia, agriculture, boating/cottagers, downstream interests, environmental groups, fishing, general public, health, industry, labor, municipalities, native people
43. St. Lawrence River (U.S.) (New York)	Massena RAP Citizen's Advisory Committee (CAC)	academia/education, agency representation, agriculture, appointed official, civic groups, environmental groups, economic/business, industry, labor, local elected officials, native people, sportsmen

the future for their area's natural resources. Great Lakes United, a coalition of citizens and citizens' organizations, has shown that citizens are eager to design their own solutions to problems (GLU 1990), because they want to change the image of their Area of Concern and make greater use of it.

Thus, meaningful public participation in these institutional structures (i.e. stakeholders' groups, basin committees, citizens' committees, public advisory committees, etc.) has brought local ownership of RAPs. These locally-based institutional structures also create in community-wide responsibility for the success or failure in the decisionmaking and remediation processes. It will undoubtedly reap additional benefits and will provide a model for governments and citizens to use in developing lakewide management plans to restore impaired beneficial uses in the open waters of the Great Lakes Basin Ecosystem. No longer is the future of an Area of Concern decided in some distant boardroom, behind closed doors; rather, the process and procedures for environmental protection have been opened fully and, once opened, cannot be closed.

Adequate Information and Research

Where there has been a long-term commitment to research on a particular Area of Concern, there is a large and relatively extensive data base that has proven invaluable in comprehensively identifying the ecosystem problems, use impairments and their causes (Hartig 1988). The availability of such natural science data and information essentially "jump-starts" the RAP development process. For example, the Hamilton Harbour RAP had immediate and direct benefit from research conducted by the Ontario Ministry of the Environment and at the Canada Centre for Inland Waters in Burlington, Ontario (a large federal research institute located on the harbor). The first step in the development of the RAP in 1985 was to prepare a technical summary of the numerous investigations on the harbor (IJC 1987a). Both the extent and completeness of the data base and the accessibility of committed researchers were major advantages to the RAP writing team.

In Green Bay, Wisconsin the RAP was based on a broad data and information base of previous academic studies and government planning initiatives (WDNR 1987). In addition, a long-standing core group of university scientists and government engineers provided the necessary leadership and technical knowledge to help synthesize the extensive data base. Likewise, for the Bay of Quinte RAP, the technical writing team benefitted from a wealth of background data collected as part of an ongoing, 20-year, multidisciplinary study of the bay (i.e. Project Quinte). The technical writing team used modeling techniques to synthesize the data and evaluate remedial options in a "Time to Decide" report (IJC 1989a).

In these cases and others, the availability of adequate data bases allows for a comprehensive review as well as the identification of previously unrecognized problems. Adequate knowledge and understanding of the issues facing the Area of Concern helps stakeholders recognize alternate approaches to resolving problems and advantages of sequencing remedial actions that are based on this knowledge of interactions within the ecosystem.

Effective Use of Existing Tools

In areas where governments and individuals have effectively used existing regulatory and resource management tools, the RAPs have progressed the furthest. It is frequently argued that the greatest short-term progress is achieved when the effectiveness and coordination of existing state, provincial and federal pollution control programs is improved, and the degree to which government priorities and policies are accountable to the publics in Areas of Concern is maximized (Munton 1988). Every effort must be made to enforce existing laws and policies. Indeed, there are numerous tools available to restore Areas of Concern. The challenge is to use these tools effectively to address environmental priorities in Areas of Concern.

Examples of effective use of existing tools include the following: use of the U.S. Superfund program to assess and remediate contaminated sediments in Ashtabula River (Ohio) and Waukegan Harbor (Illinois); use of Ontario's control orders to reduce contaminant inputs at source in Hamilton Harbour (Ontario); use of Michigan's Act 307 program to assess and remediate contaminated sediments in the Kalamazoo River; and use of Wisconsin's Priority Watershed Program to provide financial incentives for nonpoint source pollution controls in the Milwaukee and Fox River watersheds (IJC 1989a). These examples are helping to build a record of success in

Areas of Concern and sustain public interest and support in the RAP process (Hartig et al. 1991). It should be recognized, however, that use of these tools represents only the first essential step in remediation; often these tools fall short in achieving long-term goals.

RAPs have driven existing pollution control programs further and faster than could have been expected. Because of the high visibility of RAPs in the local community and the ongoing commitment to public participation, greater emphasis has been placed on taking remedial actions. Timetables for implementation have been compressed and entirely new programs have been introduced (e.g. closed-loop systems in industry; Hartig and Zarull, 1991). However, a reluctance to identify actions taken as specifically stemming from or influenced by RAPs, remains. It is not clear whether this is to avoid "loss of credit" or avoid entanglement in a process that may sometimes be perceived as delaying—rather than facilitating—remediation.

The IJC's Great Lakes Water Quality Board recognizes that the development and implementation of RAPs is a two-track process: 1) identification and implementation of remedial actions that are necessary in the near-term to resolve immediate problems, and 2) investigation and continued planning to identify and implement remedial actions to fully restore all impaired beneficial uses (IJC 1989b). Therefore, remedial actions are important milestones in the process. The IJC and its Great Lakes Water Quality Board have concluded that a schedule of key remedial action steps must be identified and achieved in order to measure progress and sustain this iterative process (IJC 1989b). Indeed, it is essential that a record of success is built to keep the momentum going on RAPs.

Political Commitment and Support

The concept of RAPs was developed by the Great Lakes Water Quality Board of the International Joint Commission and first articulated as a recommendation in its 1985 biennial report (IJC 1985). As a result, the eight Great Lakes states and the Province of Ontario committed themselves to developing and implementing a RAP to restore impaired beneficial uses in each Area of Concern within their political boundaries. These political commitments gave birth to the RAP program and were important first steps. Proof of this

commitment to RAPs has been the allocation of government resources and the personal involvement of politicians.

It is well accepted that the highest levels of management in government agencies must be genuinely committed to RAPs and demonstrate that commitment through the allocation of resources. One
good example is in Ohio, where the Governor and the Director of
the Ohio Environmental Protection Agency have both supported
Great Lakes restoration through RAPs and allocated: staff to facilitate RAP development; financial resources to perform studies that
fill data gaps; and human and financial resources to support public
participation in RAP development through advisory committees.
In addition, the Ohio Environmental Protection Agency has actively
pursued and acquired financial support for RAPs from local governments and foundations. Adequate linkages between traditionally
separate programs such as economic development, recreation, environmental protection and natural resource management have been
ensured through this political commitment and involvement.

Another good example is in Ontario, where the Ministry of the
Environment and top provincial management have incorporated
RAPs into program plans, thereby making them eligible for human
and financial resource support. Through a formal agreement between provincial and federal governments (i.e. Canada-Ontario
Agreement), resources have been allocated to support RAP development, to investigate and answer specific questions, and to facilitate
public consultation through public advisory committees and stakeholders groups. Such continued leadership and direction must come
from the highest levels of state, provincial and federal governments
and must be consistently manifested through allocation of necessary resources. Equally important as government leadership is local
ownership. Governments can provide the essential financial incentives and technical assistance that will allow local stakeholders to
take ownership of their RAPs, since the local community is most
capable and willing to plan for and ensure their own future.

Broadening the Approach from Pollution Abatement to
Ecosystem Management

RAPs have evolved from water pollution control plans to integrated
resource management plans (IJC 1989b). The IJC and its Water Qual-

ity Board view this as consistent with the adoption of an ecosystem approach called for in the Great Lakes Water Quality Agreement. Indeed, at an IJC workshop on progress and prospects of RAPs, RAP coordinators, water quality managers, researchers, members of citizens' groups and representatives of the public at large concluded that RAPs are the best tool available to integrate the diverse principles of the Great Lakes Water Quality Agreement and implement an ecosystem approach (Hartig et al. 1991). RAPs have created a more holistic way to deal with society's problems of resource management within watersheds or ecosystems.

This broadened ecosystem management structure is essential to address the complex toxic substances problems that continue to plague Areas of Concern despite extensive histories of regulatory pollution control programs. Without such holistic thinking and diverse involvement in decisionmaking, the Great Lakes community cannot address the real causes of toxic substances problems. Such a fundamental change might best be described as a step toward *"grassroots ecological democracy,"* where the realities of compartmental resource management unveil and an opportunity is created for a new order. While governments may not reorganize entire departments in response to the ecosystem approach, they must effectively integrate actions between all departments to successfully develop and implement a RAP program.

Recognition of and Commitment to a Long-Term Process

The degree and extent of toxic substances contamination in Areas of Concern has taken decades to create. The ubiquitous nature of contaminated sediments and hazardous waste sites represents the legacy of many years of release and disposal of contaminants into the environment. Therefore, a long-term commitment to mitigate the effects of toxic substances is required.

As communities have recognized that there is "no quick fix" for toxic substances problems, they are committing to a long-term process and in so doing have increased the probability for success in the RAP program. In Green Bay, Wisconsin and Rouge River, Michigan, initial goals were established in 1985 for restoring impaired beneficial uses by the year 2000 and 2005, respectively. These dates are realistic, when one considers the multiple problems, diverse sources and complex solutions. Long-term commitments also provide enough time for governments and stakeholders to undertake a systematic and comprehensive approach to identifying problems, setting goals, evaluating remedial options, choosing preferred remedial actions and securing resources for implementation. The public then is not misled into thinking that society can solve toxic substances problems in a few short years.

Problems in Areas of Concern are as much societal issues as they are environmental problems. Solving these problems through RAPs is a time consuming process of human interaction and technical application, as well as social and political dedication. Fully restoring Areas of Concern may be a long way off, but many milestones can be celebrated along the way to ensure a sense of accomplishment and continued commitment.

Purpose of This Book

This book presents 11 case studies of real, practical experiences in attempting to break down institutional barriers and implement the ecosystem approach through the development and implementation of RAPs. The case studies were chosen to illustrate a broad range of problems, solutions, and experiences from most of the Great Lakes states and the Province of Ontario. They also provide specific examples of what has been accomplished since the initial political commitments for RAPs were made in 1985.

Each of the 11 case studies is at a different stage of RAP development and implementation. The first three case studies—Green Bay, Hamilton Harbour, Rouge River—are furthest along in the RAP process. They are good examples of effective use of public participation and RAP institutional structures to implement the ecosystem approach, but stand out because they have moved into the implementation phase. Each demonstrates how the RAP process can be used as a catalyst to accelerate implementation and the value of public participation in sustaining project momentum.

Areas of Concern at an intermediate stage in RAP development include the Buffalo River, Ashtabula River, Milwaukee Estuary, Bay of Quinte and Rochester Embayment. These case studies also provide good examples of effective use of public participation and RAP institutional structures to implement the ecosystem approach, and thus a sense of local ownership of the RAP has developed in each of these areas. Rochester Embayment and Milwaukee Estuary exemplify effective watershed planning; the Bay of Quinte RAP demonstrates the value of science and effective use of modeling; and the Buffalo River RAP demonstrates effective constituency-building. The Ashtabula River RAP shows the importance and benefits of involving key politicians and obtaining upper-level management support from government.

The last three case studies on Grand Calumet River/Indiana Harbor Ship Canal, Waukegan Harbor and Kalamazoo River are all early in the RAP development phase. Kalamazoo River and Waukegan Harbor are good examples of cleaning up the most obvious problems with available tools. Grand Calumet River/Indiana Harbor Ship Canal demonstrates the use of existing tools and the important role of federal and state environmental enforcement. The book concludes with a chapter on "Keystones for Success." Keystones are, in our opinion, critical elements for continued and future success in the RAP process, and are presented to help communication, cross fertilization of ideas and, hopefully, provide insight for future directions.

Through these examples, this book celebrates the small and large victories being experienced through RAPs and encourages a spirit of partnership in restoring the integrity of the Great Lakes Basin Ecosystem. Although many problems in Areas of Concern are the same, the approaches and solutions may vary significantly. Therefore, it is important that we learn from each other's experiences and recognize our interdependency with each other and with our ecosystem.

REFERENCES

Canada and the United States. 1987. *Protocol to the Great Lakes Water Quality Agreement*. Windsor, Ontario, Canada.

Christie, W.J., M. Becker, J.W. Cowden, and J.R. Vallentyne. 1986. Managing the Great Lakes as a Home. *J. Great Lakes Res.* 12:2–17.

Colborn, T.E., A. Davidson, S.N. Greau, R.A. Hodge, C.I. Jackson, and R.A. Liroff. 1990. Great Lakes-Great Legacy? *The Conservation Foundation (Wash. D.C.) and The Institute for Research on Public Policy*. Ottawa, Ontario.

Francis, G.R., J.J. Magnuson, H.A. Regier, and D.R. Talhelm. 1979. Rehabilitating Great Lakes ecosystems. *Great Lakes Fishery Commission Tech. Rep.* 37, Ann Arbor, Michigan.

Great Lakes United (GLU). 1990. RAP Revival: A Citizen's Agenda for RAPs. Buffalo, New York.

Harris, H.J., P.E. Sager, S. Richman, V.A. Harris, and C.J. Yarbrough. 1987. Coupling ecosystem science with management: a Great Lakes perspective from Green Bay, Lake Michigan, USA. *Env. Man.* 11:619–625.

Hartig, J.H. 1988. Remedial action plans: A Great Lakes program whose time has come! The Great Lakes: Living with North America's Inland Waters, D.H. Hickcox (ed.), *Amer. Wat. Res. Assn.* TPS-88-3:45–51.

Hartig, J.H. and R.L. Thomas. 1988. Development of plans to restore degraded areas of the Great Lakes. *Env. Man.* 12:327–347.

Hartig, J.H. and J.R. Vallentyne. 1989. Use of an ecosystem approach to restore degraded areas of the Great Lakes. *AMBIO* 18:423–428.

Hartig, J.H. and P.D. Hartig. 1990. Remedial action plans: An opportunity to implement sustainable development at the grassroots level in the Great Lakes Basin. *Alternatives* 17:26–31.

Hartig, J.H., L. Lovett Doust, and P. Seidl. 1990. Successes and challenges in developing and implementing remedial action plans to restore degraded areas of the Great Lakes. International and Transboundary Water Resources Issues, J.E. FitzGibbon (ed.), *Amer. Wat. Res. Assn.* TPS 90–1:269–278.

Hartig, J.H. and M.A. Zarull. 1991. Methods of restoring degraded areas in the Great Lakes. *Rev. Env. Cont. Tox.* 117:127–154.

Hartig, J.H., L. Lovett Doust, M.A. Zarull, S. Leppard, L.A. New, S. Skavroneck, T. Eder, T. Coape-Arnold, and G. Daniel. 1991. Overcoming obstacles in Great Lakes remedial action plans. *Int. Env. Affairs.* 3:91–107.

International Joint Commission (IJC). 1978. The ecosystem approach. *Great Lakes Science Advisory Board*, Windsor, Ontario, Canada.

IJC. 1985. Report on Great Lakes Water Quality. *Great Lakes Water Quality Board*, Windsor, Ontario, Canada.

IJC. 1987a. Progress in developing remedial action plans for Areas of Concern in the Great Lakes basin. Appendix A: 1987 Report on Great Lakes Water Quality. *Great Lakes Water Quality Board*, Windsor, Ontario, Canada.

IJC. 1987b. Report on Great Lakes Water Quality. *Great Lakes Water Quality Board*, Windsor, Ontario, Canada.

IJC. 1989a. *Fourth Biennial Report*. Windsor, Ontario, Canada.

IJC. 1989b. Report on Great Lakes Water Quality. *Great Lakes Water Quality Board*, Windsor, Ontario, Canada.

Munton, D. 1988. *Toward a more accountable process: The Royal Society-National Research Council Report*. Perspectives on Ecosystem Management for the Great Lakes, L.K. Caldwell (ed.), pp. 299–317. State Univ. of New York Press, Albany, New York.

Thomas, R.L. and J.H. Hartig. 1989. Lake case study: the Great Lakes. Global Freshwater Quality: A First Assessment, M. Meybeck, D. Chapman and R. Helmer (ed.), *World Health Organization*, Oxford, UK: Basil Blackwell Inc.

Thomas, R.L., J.R. Vallentyne, K. Ogilvie, and J.D. Kingham. 1988. *The ecosystems approach: A strategy for the management of renewable resources in the Great*

Lakes Basin. Perspectives on Ecosystem Management for the Great Lakes Basin, L.K. Caldwell (ed.), pp. 31–57. State Univ. of New York Press, Albany, New York.

Vallentyne, J.R. 1974. The algal bowl. Department of the Environment, Fisheries and Marine Service. *Spec. Publ.* 22. Ottawa, Ontario, Canada.

Wisconsin Dept. of Natural Resources (Wisconsin DNR). 1987. *Lower Green Bay Remedial Action Plan.* Publ. WR-175-87. Madison, Wisconsin.

Chapter 2

From Plan to Action: The Green Bay Experience

Victoria A. Harris

"We in Wisconsin are proud of our record of working together to protect our state's waters. This concept of teamwork continues through our RAP projects, as evidenced by the success of the Green Bay RAP. Each RAP has created partnerships of local governmental, business and community leaders. These new partnerships are the foundation upon which implementation programs will be built."

Governor Tommy Thompson
State of Wisconsin

Introduction

The lower Green Bay and the Fox River comprise one of 43 Areas of Concern identified by the International Joint Commission (IJC) as needing a remedial action plan (RAP) to correct ongoing water quality problems and restore beneficial uses. A RAP was prepared by the Wisconsin Department of Natural Resources (DNR) from 1985 to 1987 with cooperation from other agencies, researchers, and citizens of northeastern Wisconsin. The RAP was adopted by the Wisconsin DNR and the Governor's office in February 1988 and is now in its third year of implementation.

Because the Lower Green Bay and Fox River RAP was one of the first RAPs to be completed and the first to be accepted by the International Joint Commission and its Water Quality Board, it has served as an experiment and model for RAP development and implementation processes and the ecosystem approach to planning and management. Many successes and challenges have occurred along the way, and this chapter shares some of the strengths and weaknesses of the approaches taken in the Lower Green Bay and Fox River RAP. The Area of Concern, the problems addressed, the planning process, key actions of the plan and the implementation approach are also described, and seven critical elements required for RAP implementation are suggested.

The Area of Concern

The Lower Green Bay and Fox River Area of Concern is located at the head of the world's largest freshwater estuary, where the Fox River empties into southern Green Bay on Lake Michigan. The Area of Concern encompasses the last 11.2 km (seven miles) of the Fox River from the DePere Dam to its mouth, and the "inner bay" out to Longtail Point on the west shore and Point Au Sable on the east shore (Figure 4). The area is wholly within the State of Wisconsin.

Green Bay is defined as an estuary because of the strong thermal, chemical, and trophic gradients that exist along the axis of the bay between the Fox River and Lake Michigan. The inner bay is at most 3–4 m deep and not stratified. It is rapidly flushed, hypereutrophic, highly turbid, and contains mostly Fox River water, which has high dissolved and particulate anthropogenic substances. In contrast, the northern bay is 30–35 m deep, strongly stratified, oligotrophic and resembles Lake Michigan in its water quality and clarity. The west shore of the lower bay has low lying sandy plains and coastal wetlands, while the east shore is characterized by rock outcroppings, clay soils, and residential development.

The once wild Fox River has been tamed by 17 locks and dams that trap sediments and pollutants. Below the DePere Dam, the river has a low gradient and is often channelized and flanked by intense urban and industrial development. While water quality problems are most acute in the Area of Concern, the sources of many problems lie upstream in the large, 17,200 km^2 (6,641 square miles) drainage basin.

The drainage basin includes watersheds of the upper Fox River, Wolf River, Winnebago Pool Lakes and Lower Fox River. Much of the land surface has been altered for agricultural, industrial, or residential uses. About 750,000 people reside in the basin—over half in the heavily industrialized Fox River valley. Approximately 60 municipalities and over 100 industries discharge into the basin. The 64 km (40 mile) reach of the Lower Fox River receives the discharges of 13 pulp and paper mills, the largest concentration of the paper industry in the world.

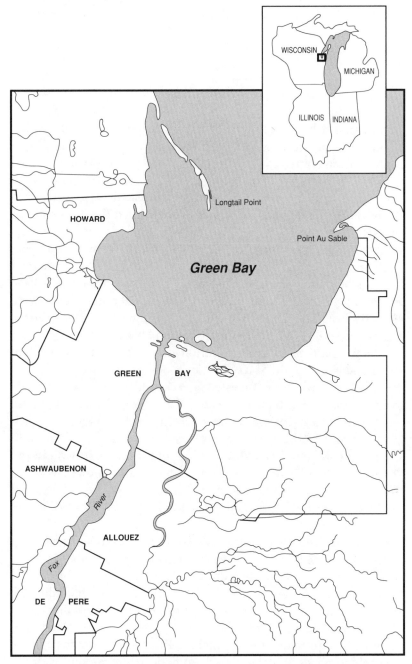

Fig. 4. Location of the Lower Green Bay and Fox River Area of Concern; inset shows area of enlargement. (Source: Bay Lake Regional Planning Commission.)

The Problem

Lower Green Bay and the Fox River have a long history of resource use and abuse (Harris et al. 1987a, Smith et al. 1988). Intensive commercial fishing and market hunting depleted fish and wildlife populations in the late 1800s. Following the lumber era, cleared lands were converted to agricultural uses and the river valley turned to industry. With rapid population growth, communities and industries discharged increasing volumes of untreated wastewater into the bay. Oxygen-demanding wastes overwhelmed the river and bay ecosystem by the 1920s and fish kills became commonplace. Surveys of benthic organisms in the 1920s and 1950s indicated some areas of the river and bay were devoid of invertebrates. Commercial harvests of yellow perch reached an all-time low in the mid-1960s. Native stocks of lake trout, sturgeon, herring and muskellunge were virtually lost. The Green Bay public swimming beach was permanently closed in 1943 due to unsafe bacteria levels.

Effective efforts to curb pollution didn't begin until the late 1960s and early 1970s. Since then, over 300 million public and private dollars have been invested in secondary wastewater treatment to meet a wasteload allocation for oxygen-demanding waste (i.e. biochemical oxygen demand). As a result, oxygen and fish have returned to the river and lower bay (Ball et al. 1985). Over 30 species of fish were found during 1985 mid-summer surveys in the Fox River below DePere. Improving water quality allowed Wisconsin DNR to stock walleye in an effort to restore the fishery, and the lower river now boasts one of the best walleye fisheries in the nation.

Wisconsin fish consumption advisories still warn against eating walleye from the Lower Fox River and Green Bay, as well as 11 other species of fish and mallard ducks due to PCB (polychlorinated biphenyl) contamination. Over 100 toxic chemicals, including 39 U.S. Environmental Protection Agency priority pollutants, have been detected in Fox River discharges, water, sediments and fish. Chemicals of primary concern include PCBs, some dioxins and furans, chlorophenols, polynuclear aromatic hydrocarbons, volatile organic compounds, PCB substitute compounds (e.g. isopropylbiphenyl) mercury, lead, pesticides (DDT, DDE) and ammonia. In the early 1980s, populations of fish-eating birds exhibited reproductive problems, behavioral disorders and wasting syndromes that have been symptomatically linked with chemicals like PCBs and dioxins (Kubiak et al. 1989).

The lower bay is hypereutrophic and receives an estimated annual load of more than 500 metric tons of total phosphorus. Approximately one-third is contributed by municipal and industrial discharges, while the remaining two-thirds enters from watershed runoff, including urban and rural nonpoint sources of pollution (Harris and Christie, 1987). The Area of Concern receives a total average annual loading of 90,000 metric tons of sediments, nutrients and suspended solids.

While toxic substances have captured the most public and agency attention, more impaired uses are associated with hypereutrophication and sedimentation in the Area of Concern than with toxic contamination (Figure 5). Excessive nutrient loading has altered the species composition and size distribution of phytoplankton and enhanced the growth of blue-green algae (Richman et al. 1984). Correspondingly, zooplankton size distribution, grazing rates and consumptive capacity have also shifted. Much of the primary productivity is not efficiently used through the pelagic food chain (Sager, in press) and is diverted to the detrital food chain where it depletes sediment oxygen, contributes substantially to ammonia toxicity in the sediments, and degrades benthic populations. Recent bioassays show sediment toxicity in the Area of Concern is due mainly to ammonia generated from the high rates of decomposition of organic matter (algae) and not to other toxic substances (Ankley et al. 1990).

Sedimentation and suspended solids fill navigation channels, destroy spawning habitats and impede light penetration. About 250,000 m^3 of sediment must be dredged annually from the harbor and entrance channel. While dredging is not restricted in the Area of Concern, disposal of these sediments—all contaminated to some degree—has been highly controversial. Both suspended sediments and algae combine to create turbid water conditions that prevent sufficient light penetration to support the re-establishment of submerged macrophytes. One of the most striking ecological changes in the lower bay ecosystem has been the loss of submerged aquatic vegetation needed for fish spawning and as food for waterfowl.

Swimming beaches remain closed in the lower bay because waters do not meet safety standards for water clarity (1.3 m Secchi

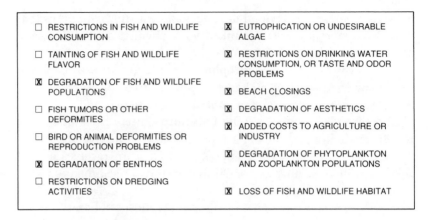

☐ RESTRICTIONS IN FISH AND WILDLIFE CONSUMPTION	☒ EUTROPHICATION OR UNDESIRABLE ALGAE
☐ TAINTING OF FISH AND WILDLIFE FLAVOR	☒ RESTRICTIONS ON DRINKING WATER CONSUMPTION, OR TASTE AND ODOR PROBLEMS
☒ DEGRADATION OF FISH AND WILDLIFE POPULATIONS	☒ BEACH CLOSINGS
☐ FISH TUMORS OR OTHER DEFORMITIES	☒ DEGRADATION OF AESTHETICS
☐ BIRD OR ANIMAL DEFORMITIES OR REPRODUCTION PROBLEMS	☒ ADDED COSTS TO AGRICULTURE OR INDUSTRY
☒ DEGRADATION OF BENTHOS	☒ DEGRADATION OF PHYTOPLANKTON AND ZOOPLANKTON POPULATIONS
☐ RESTRICTIONS ON DREDGING ACTIVITIES	☒ LOSS OF FISH AND WILDLIFE HABITAT

Fig. 5. Impaired uses due to nutrient and sediment loading to the Lower Green Bay and Fox River Area of Concern

disk) and indicator bacteria levels also frequently exceed public safety standards. Algal "blooms" also detract from other recreational uses and the aesthetic character of the bay, increase costs for water treatment prior to industrial use, and contribute to restrictions on drinking water use.

Nearly 90 percent of the coastal marshes have been lost in the Area of Concern due to dredge spoil disposal, land filling, industrial and residential development, and recent high lake levels. Waterfowl use of the bay has declined over the past 50 years, as have colonial waterbird nesting and furbearer populations. Native species of certain predator fish (walleye, muskellunge, northern pike and bass) have been largely replaced with pollution tolerant bottom-feeding fish (carp, white sucker, catfish and bullhead). Exotic species (carp, sea lamprey, alewife, white perch and gizzard shad) further threaten remaining native fish stocks.

Beyond these problems, responsibility for management of lands and waters in the Fox River/Green Bay watershed has been spread among a multitude of agencies and governments, complicating the restoration process. These agencies often have disparate objectives and policies, and no mandate or mechanism for coordinated ecosystem management existed. Until the RAP program began, actions were biased towards the objectives of individual institutions, each operating more or less independently of one another, and there were few incentives to coordinate efforts. In fact, there may have even been rewards for noncooperation.

The Remedial Action Planning Process

Wisconsin DNR prepared the RAP over a period of two years (1985–1987) with assistance from other agencies and stakeholder groups (Figure 6). More than 75 people participated on a Citizens Advisory Committee (CAC) and four Technical Advisory Committees (TAC).

The CAC was established early in the planning process and remained active until the RAP was adopted and an Implementation Committee was created. The CAC represented the diverse interests and perspectives of business, industry, recreation, conservation and environmental groups, shoreline residents, local government and federal, state and local agencies. The CAC helped to identify problems that should be addressed by the plan and developed goal statements for the "desired future state" of the river and bay. It reviewed the work of the TACs and provided advice on the selection of preferred remedial strategies. It also participated in public education and consultation by sponsoring informational meetings, workshops, field trips, and a photo and poster contest for school groups.

The TACs addressed the four primary problem areas identified by Wisconsin DNR staff, scientists and the CAC: toxic substances; nutrients and eutrophication; biota and habitat; and institutional/recreational concerns. TAC members included researchers, resource managers, local experts and one or two liaison members from the CAC. The TACs analyzed problems and prepared reports to define use impairments in the Area of Concern, identify ecosystem-oriented goals and objectives, and recommend alternative remedial strategies to restore beneficial uses. Following public and agency reviews, the TAC reports formed the technical basis for the RAP.

Other citizen input was obtained from two widely distributed questionnaires on present and future resource uses, perceptions of water quality and desired improvements. Citizen comments on draft documents were received through several public meetings and hearings.

The final plan was submitted to the IJC for review in October 1987 and was approved by Wisconsin DNR in February 1988. The

I. Developing Scope of Study
  • Problems plan should address
  • Objectives for plan

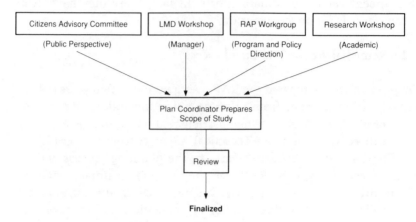

II. Preparation of Plan

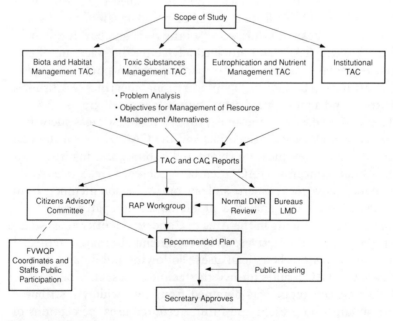

DNR = Department of Natural Resources
CAC = Citizens Advisory Committee
LMD = Lake Michigan District
TAC = Technical Advisory Committee
FVWQP = Fox Valley Water Quality Planning Agency

Fig. 6. Planning process for Lower Green Bay Remedial Action Plan

RAP was subsequently adopted by the Governor's office as a part of the state's Water Quality Management Plan and accepted by the IJC and its Great Lakes Water Quality Board.

Planning Precursors

Several catalysts helped to make the RAP happen. The RAP benefited greatly from previous water quality planning and rehabilitation efforts, including: an allocation for biochemical oxygen demand (BOD) for the Fox River; major public and private expenditures to implement the wasteload allocation and resultant water quality improvements; and renewed public interest and heightened expectations for public use of the river and bay as a result of these actions. Many individuals involved on public advisory committees for the wasteload allocation became actively involved in the RAP process.

The University of Wisconsin Sea Grant Institute and other scientists also contributed a wealth of technical information to the RAP TACs. The Sea Grant Institute had reinstated a separate Green Bay Program in 1978 to improve knowledge and management of the lower bay ecosystem (Harris and Garsow, 1978). This program has continued for more than 12 years and provides valuable insight into the changing structure and function of the lower bay ecosystem.

The Sea Grant research program on Green Bay also attracted the attention of the Great Lakes Fishery Commission, which was looking for candidate sites to conduct feasibility studies of Great Lakes Ecosystem Rehabilitation (GLER). The Fishery Commission identified two Great Lakes ecosystems with pressing rehabilitation needs: Green Bay (Lake Michigan) and the Bay of Quinte (Lake Ontario). In both areas, factors affecting the ecosystem had been identified and the attention of scientists, resource managers and citizens was actively focused on reversing the deterioration of these aquatic systems. The GLER study concluded that: 1) comprehensive ecosystem rehabilitation strategies for the Great Lakes are, in general, feasible to develop; 2) strategies should be initiated first for smaller ecosystems such as bays and harbors, and tailored to the particular conditions and stresses impacting particular areas; and 3) strategies should address measures to alleviate the key stresses affecting the aquatic ecosystem and specify rehabilitation objectives in terms of a consistent set of ecological indices and the conditions needed to sustain them (Francis et al. 1979).

GLER workshops and seminars held in 1979 and 1980 resulted in a "Rehabilitative Prospectus" for Green Bay (Harris et al. 1982). The plan was never officially adopted by state or federal agencies; however, the GLER project spawned a coalition of agencies and citizen interests, called "Future of the Bay," which attempted to promote an ecosystem approach and coordinated management for Lower Green Bay in the early 1980s. The prospectus would later provide a starting point for the RAP.

When RAP planning began in 1985, the U.S. Environmental Protection Agency, University of Wisconsin-Green Bay and Wisconsin DNR cosponsored a workshop for Great Lakes scientists and local resource managers to identify a set of ecological properties or indices that would guide restoration and measure the integrity of the lower bay ecosystem. These included carbon transfer efficiencies between trophic levels, presence or absence of sensitive benthic organisms, and body burden of toxic chemicals, growth rates, and fecundity of top predator fish such as walleye. These properties characterized Lower Green Bay in particular and the Great Lakes ecosystem in general, and were used to assess the state of ecosystem impairment and/or restoration (Harris et al. 1987b). Since IJC criteria for listing/delisting Great Lakes Areas of Concern had not yet been developed, these indices later helped to develop ecosystem-oriented objectives for the RAP.

The Remedial Action Plan

The underlying premise of the RAP is to move beyond the limitations of previous conventional point source pollution controls of the 1970s and 1980s to integrated resource management in the 1990s. The Lower Green Bay and Fox River RAP contains 16 key actions for ecosystem rehabilitation and 120 specific recommendations. Each recommendation identifies a series of action steps, target dates, estimated costs, and a list of implementing agencies. Detailed implementation strategies and a designated lead agency were identified later in the RAP implementation process.

The key priority actions and some of their recommendations include:

- Reduce total phosphorus loading to the Area of Concern by 40–50 percent through water quality standards for phosphorus; lower effluent phosphorus limits for municipal and industrial discharges; and implement 11 to 20 comprehensive nonpoint source watershed management projects and regulations for construction erosion control, stormwater management, animal waste management and streambank pasturing.
- Reduce sediment and suspended solids inputs through the same nonpoint source pollution measures recommended for phosphorus.
- Eliminate the toxicity of point source discharges by adopting water quality criteria and standards for toxic substances and requirements for effluent limits and monitoring in discharge permits.
- Reduce inputs of toxic chemicals from contaminated sediments.
- Continue control of oxygen-demanding wastes through a wasteload allocation for municipal and industrial discharges.
- Create an institutional structure to coordinate RAP implementation.
- Increase public awareness, participation and support for river and bay restoration.

There were several strengths of the Lower Green Bay RAP process. For example, all stakeholders were committed to citizen participation throughout the RAP process, broad representation of stakeholder groups on the advisory committees, and a vigilance in maintaining an ecosystem perspective. There was a strong tendency for specific stakeholder groups to become single-issue oriented, since most have specific concerns or reasons for participating. Thus an important part of the RAP process was continued education on the interrelatedness of problems and potential solutions. Participation by scientists and the previous research made it possible to develop integrated resource objectives that, for example, link water clarity objectives for macrophyte re-establishment with phosphorus-loading objectives, and waterfowl use objectives with the abundance of fingernail clams.

Weaknesses in the RAP process also existed. Insufficient technical and financial information was available to specify some remedial actions, particularly for contaminated sediments, and the large number of plan recommendations required 120 separate implementation strategies. Some implementation strategies were incomplete for plan recommendations. Generally, it was easier to reach consensus on what to do than on how to do it. In addition, implementation of recommendations often was assigned to a group of agencies or units of government without designating a lead responsibility. Plan recommendations were organized by key action category rather than lead agency, and thus it was difficult for an agency to quickly determine its responsibilities under the plan.

The Implementation Process

As the primary water quality planning and management agency for the state, Wisconsin DNR committed to promote plan implementation, maintain the strong points of the planning process into the implementation phase and improve on the weak points. Implementation planning began several months before the RAP was written; yet even that was not early enough to obtain the institutional and political consensus necessary to create a new institution for plan implementation. Therefore, an interim RAP implementation structure and process was defined during RAP development and has since continued to evolve.

The Institutional TAC identified several alternative management structures for RAP implementation, including: 1) a coalition of stakeholders and implementing agencies; 2) a coordinating council created by state legislation and appointed by the governor to advise governments on RAP implementation and to review their work programs; and 3) a basin authority with the power to tax and enact regulations to implement the RAP. The TAC felt that a structure with the greatest financial and regulatory powers (such as a basin authority) would be most effective, but questioned the political feasibility of creating another layer of government. Such a special purpose unit of government would need to be endorsed by each local unit of government in the basin, and a previous attempt to create a basin authority for the Fox River watershed had failed.

Implementation Committee

Instead, the TAC and CAC recommended that the Wisconsin DNR create a local Implementation Committee for an interim period of two years to initiate plan implementation and to further evaluate institutional options for long-term plan implementation. The Implementation Committee is a consortium of stakeholder groups that advises Wisconsin DNR, local governments, and others on RAP implementation. Its 30 members represent lead agencies, local governments, the state legislature, business, industry, environmental groups, conservation groups and citizens with recognized understanding and influence in the Area of Concern. While Wisconsin DNR appointed specific members to the Implementation Committee, the former CAC and Institutional TAC provided the membership design. It reflects the organizations and institutions whose cooperation was considered essential for plan implementation. About one-third of the membership of the CAC was extended to the Implementation Committee and this provided important continuity in public participation from planning to implementation.

The Green Bay RAP Implementation Committee is assisted by a steering committee and six advisory committees, each working on implementation strategies for different key actions and recommendations (Figure 7). Over 100 individuals representing more than 60 organizations participate on the committees.

This large structure has benefits and drawbacks. It brings together diverse viewpoints, talents and experience, but this diversity can lead to both innovation and conflict. Broad stakeholder representation creates a greater awareness and a sense of ownership in the plan, which contribute directly to implementation of actions by participating organizations. However, the complex committee structure and bimonthly meetings make it difficult to move quickly on issues or opportunities. Members are largely upper level administrators with little time for extra meetings or outside work. A tremendous workload is associated with coordinating committee activities: the equivalent of three full-time staff members are required to support committee work, which is provided by Wisconsin DNR, University of Wisconsin-Green Bay, Bay-Lake Regional Planning Commission and by committee members and their agencies. Staffing and operating the Implementation Committee structure

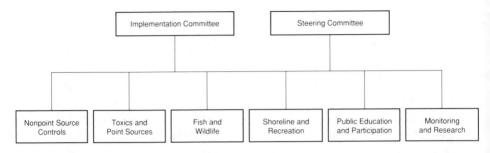

Fig. 7. Lower Green Bay Remedial Action Plan Interim Implementation Committee Structure

costs about $150,000–$200,000 per year. Funding is provided by Wisconsin DNR and grants from the Wisconsin Coastal Management Program and Brown and Outagamie Counties.

The charge to the Implementation Committee was developed by the Wisconsin DNR and modified and adopted by the committee. It is extremely broad and encompasses everything but actual implementation. The charge includes: advising the Wisconsin DNR, local governments and others on RAP implementation strategies and plan updates; setting priorities and timetables; seeking funds; promoting cooperative efforts and an ecosystem approach to management; coordinating activities and information sharing; conflict resolution; reviewing and reporting on implementation progress; and public education and participation.

During the first year the Implementation and Advisory Committees focused on getting organized and becoming informed, seeking funding and staff support, establishing priorities and work plans, refining implementation strategies and costs, identifying responsible "lead agencies," and developing a public information and participation plan. These actions were time consuming but necessary first steps to make the RAP management structure operational. A "RAP-Track" information system was developed to track the status of each RAP recommendation and to prepare the First Annual Progress Report.

During the second year, the Implementation Committee developed a more direct and structured approach to plan implementation, building on adopted RAP priorities and strategies. The RAP was "repackaged" into guidebooks for designated lead agencies, which summarized the RAP's key actions and recommendations as assigned to particular organizations (i.e. regional planning commissions, municipal wastewater treatment plants and county governments) and included target dates and suggested implementation strategies for each remedial action. The Implementation Committee then formed "contact teams" to approach lead agencies and influence them to take remedial actions. The contact teams inform lead agencies of their role in the RAP, solicit written commitments for actions, identify obstacles and solutions to implementation, and assess progress through followup contacts. The process includes preparation of an annual progress report and a public hearing. Feedback from lead agency contacts and the public is used to review and revise implementation priorities and strategies. The entire process is repeated each year.

The committees also review ongoing programs and projects and write numerous letters to advise lead agencies on immediate opportunities or concerns for RAP implementation. Other committee activities include resource mapping for a geographic information system, obtaining project grants, lobbying for legislation and program resources, and developing a RAP monitoring plan. Considerable emphasis and resources have been directed toward public education and participation activities. Written materials are produced, including a quarterly newsletter, the annual progress reports, a "State of the Bay" report, and informational brochures. A speakers' bureau, slide/tape show and a teacher's guide to the RAP have been developed for use by schools and civic groups. The RAP committees cosponsor an annual River/Bay Clean Up Day, an annual Green Bay/Fox River Monitoring and Research Workshop, an annual public hearing on RAP progress and monthly "Edge of the Lake" seminars on issues related to the RAP. Committee members have assisted with informational displays at numerous public events and have sought funding for permanent RAP displays at several parks and boat landings in the Area of Concern.

Most committee activities are being continued as the RAP enters its third year of implementation. Implementation priorities and strategies are updated annually and feedback is provided to the committee and lead agency contacts. The Implementation Committee is now focusing on evaluation of the interim structure and is developing recommendations for a revised management structure to guide the RAP through completion and restoration of impaired uses in the Area of Concern.

Progress

Most remedial actions taken to date have been initiated by the organizations participating directly on the RAP committees. More effort is needed to extend participation beyond the committees and the immediate Area of Concern to upstream local governments and others who may not be aware of their impacts on downstream water quality.

Plan implementation is generally on schedule for most high priority recommendations, although many initiated projects are long term and will take years to complete. Progress has been made on all key actions and over half of the plan's 120 recommendations. A summary of the projects and funds (in thousands of dollars) contributing toward the Green Bay RAP in 1990 is shown in Table 3. Some of these actions or programs would have occurred without the RAP, although perhaps not as quickly; others are the direct result of the RAP and/or the implementation committees. Notable accomplishments to date include:

- Adoption of state water quality standards and effluent limits for toxic chemicals, including mass limits for 21 persistent, bioaccumulative substances identified by the Great Lakes Water Quality Agreement.
- Adoption of state air quality regulations that require reporting and impose emission standards on 400 toxic substances.
- A $13 million mass balance study (U.S. Environmental Protection Agency, Wisconsin DNR and University of Wisconsin Sea Grant Institute) that defines the sources, pathways and fate of four toxic substances (PCBs, Pb, Cd and Zn) in Green Bay and a PCB/sediment transport study (Wisconsin DNR) that assesses the quantity, location and movement of PCBs in the Fox River. Although the mass balance study was not intended specifically for the RAP, both studies are necessary to develop remedial strategies for contaminated sediments.
- An Implementation Committee member introduced legislation and the committee successfully lobbied the state to create an

in place pollutant management program, which provides partial support for the above studies and for a feasibility study to remove and/or treat contaminated sediments in Little Lake Butte des Morts at the head of the Fox River.

- Three large-scale priority watershed projects have been initiated that provide funding for comprehensive nonpoint source pollution management in rural and urban areas. Prior to the RAP, no priority watershed projects existed in the lower Fox River basin.
- Local governments are developing ordinances to control construction erosion and to prohibit streambank pasturing.
- Two of the largest municipal dischargers in the basin have agreed to requests by the Implementation Committee to conduct cost effectiveness analyses for additional phosphorus removal. Green Bay Metropolitan Sewerage District has voluntarily reduced its phosphorus discharge from concentrations of nearly 1 mg/L to around 0.5 mg/L.
- Several municipal wastewater treatment facilities are being upgraded to meet more stringent limitations for ammonia and chlorine discharges.
- Wisconsin DNR is stocking a Great Lakes strain of muskellunge in Lower Green Bay to increase predator fish ratios.
- Local communities have matched state funds to expand public access at boat launches, shoreline fishing spots and recreational trails along the Fox River.
- Wisconsin DNR, the Implementation Committee and local government jointly sponsored the construction of two walleye spawning beds on the Fox River and the creation of wetland habitat near the bay shore.

Challenges for RAP Implementation

RAPs present substantial challenges that must be met to successfully restore beneficial uses. The persistence of environmental problems in Areas of Concern is testimony that remedial actions will be difficult and expensive to implement. The challenges are technical, financial, and social.

While our understanding of the stresses on Great Lakes ecosystems is growing, our knowledge is insufficient in many areas, which

TABLE 3.   Lower Green Bay Remedial Action Plan (RAP) Resources Committed to RAP Projects (in thousands of dollars)

Project	Federal	State	Local	Total
Toxic Substances and				
Point Source Management				
Green Bay Mass Balance	12,146	387		12,533
PCB/Sediment Transport		798		798
Lake Butte des Morts	100	100		200
Green Bay Metro Sewerage District		15,500	53,500	69,000
Appleton Wastewater Treatment				
Plant		14,000	40,000	54,000
Industry			unknown	
Total				136,531
Nonpoint Source Pollution				
East River Watershed		25,000		25,000
East Winnebago Watershed		5,000		5,000
East River Nutrient/Pesticide	800			800
Management				
Conservation Reserve Program	1,104			1,104
Total				31,904
Fish and Wildlife Management				
Northern Pike Study		2		2
Musky Stocking		7	10	17
Create Spawning Habitat		35	9	44
Carp Barrier		5		5
Wetland Creation		8		8
Habitat (GIS) Mapping	20	74		94
Total				170
Public Access				
Fox River Trail		53	11	64
De Pere Fishing Dock		19		19
Duck Creek Boat Launch		23.5	23.5	47
Total				130
Public Education				
Displays	1.5	6.5		8
"State of the Bay"	10	10	1	21
"Baybook"	10	9		19
Public Perception Survey	25	25		50
NEWSRAP, Speakers Bureau		18	2	20
Total				118
RAP Committee Support				
Totals	14,231	61,138	93,591	168,960

can lead to disagreements or hesitation to implement costly actions. The challenge is to move ahead with confidence in implementing remedial actions while continuing to seek "better" information. Some technical questions still need to be answered: how can we cost-effectively reduce the availability of toxic substances in the sediments? how clean is clean enough? how do we remove ever smaller amounts of pollutants from our wastewater? These questions are not unique to Green Bay, and further research will be required to answer them.

Tremendous financial resources are necessary to implement RAPs. Cost estimates for the Lower Green Bay RAP range from $68 million to $640 million, and the level of funding committed to RAP projects to date (Table 3) indicates that these estimates are probably too low. Who will pay the massive costs to fully implement the plan? Current funding sources for Green Bay primarily have been local (55 percent), followed by state (36 percent) and federal sources (9 percent). Wisconsin DNR has proposed a new "water protection fee" for municipalities and industries that discharge to state waters, which would be based on the volume of water used from surface or groundwaters and would be used to fund the Great Lakes Protection Fund, Great Lakes harbors and bays cleanup and other state water management programs. Although "new" or enhanced sources of funding are vital for RAPs, resources are finite and must be shared with other important societal needs. The challenge is to redirect and sustain the resources that will be needed over the 15 to 20 years required to implement all remedial actions.

Social change also is needed to implement RAPs. This may present the greatest challenge to restoring a sustainable environment to the Great Lakes, since reducing unnecessary uses of toxic chemicals and controlling nonpoint source pollution requires changes in individual behaviors and lifestyles. The public may be willing to make such changes provided they become informed and understand the impacts of their actions, and suitable alternative products and practices become available.

Conclusions

Based on the Green Bay RAP experience, the following are suggested as critical elements for RAP implementation:

1. *Public education* must include school groups, community leaders, committee members and stakeholder groups. The ecosystem approach requires an understanding of interconnected problems and solutions, but most stakeholders are focused on specific problems or solutions. For example, so much attention has been given to toxic substances problems that many people believe the Green Bay/Fox River ecosystem can be restored simply through toxic discharge controls. While eliminating the toxicity of discharges and reducing the availability of toxic substances from contaminated sediments are high priority actions, the dominant toxicity problem in the sediments is due to ammonia primarily from algal production and decay. The greatest ecological perturbations to Lower Green Bay stem from eutrophication, sedimentation and habitat destruction. Heightened public awareness about these related problems will help us to make headway on the massive nonpoint source pollution problem.

2. *Public participation* is vital to the implementation process, although it creates risks and benefits. Heightened public expectation can lead to discontent when remedial actions take years to implement and results are not immediately apparent. There are as many views on *how* to implement as there are on *what* to implement. However, public participation in Green Bay RAP implementation has resulted in more innovative solutions and greater public support, and it has probably pushed some actions faster and farther than otherwise expected.

3. *Political will* for new policies, regulations and resources must be created through demonstrated local actions, lobbying and grassroots support. Public education and participation generate political will.

4. *Resources* must be developed or redirected through political will and lead agency participation in the RAP process. Agencies are generally more willing to commit resources to a plan they helped to establish. Smaller demonstration projects and local initiatives also can leverage funding for larger actions.

5. *Continuous analysis, flexible strategies* and an ongoing commitment to *long-range planning* are necessary to the iterative RAP process. Plans and implementation strategies need to be

periodically updated to reflect new information, address new problems and opportunities, and seek new solutions when present approaches fail.

6. *Trend monitoring and public perception surveys* provide important feedback for planning implementation and identifying ecosystem change in response to remedial actions. An understanding of public perceptions and concerns helps to ensure that public education and participation programs are effective.

7. A commitment to an *ecosystem mandate* by the lead management agencies is needed to fully restore and sustain Great Lakes ecosystems. Those organizations who participate actively in RAP implementation have vested interests or have a clear mandate to manage some aspect of the system. Those that do not participate either do not see their connectedness to the problems or do not have a sense of responsibility for action because of (in their view) narrowly defined missions and mandates. Even environmental protection agencies have disparate programs that do not always share the same objectives. New multijurisdictional agreements and institutional arrangements are needed and will require a sense of common purpose and direction. An ecosystem mandate that requires cooperation and integrates programs and objectives should be incorporated in legislation, policies and strategic plans at the federal, state, and local levels.

The Green Bay RAP is on schedule for most high priority recommendations, though many actions will take years to complete and may require new regulations, policies, resources and programs. The entire RAP process has focused more attention and resources on the Area of Concern than ever before and has moved existing programs further and faster than otherwise expected. This process must continue to evolve and expand throughout the implementation phase, and greater awareness and involvement by local governments and citizens upstream of the Area of Concern is needed to address the massive nonpoint source pollution and contaminated sediment problems. The success of the RAP also will depend on more effective institutional arrangements that include commitments for cooperative, integrated resource management and ecosystem redevelopment.

REFERENCES

Ankley, G.T., A. Katko, and J.W. Arthur. 1990. Identification of ammonia as an important sediment-associated toxicant in the lower Fox River and Green Bay, Wisconsin. *Environmental Toxicology and Chemistry.* 9:313–322.

Ball, J.R., V.A. Harris, and D. Patterson. 1985. Lower Fox River—DePere to Green Bay water quality standards review. *Wisconsin Dept. of Natural Resources, Water Resources Management,* Madison, Wisconsin.

Francis, G.R., J.J. Magnuson, H.A. Regier, and D.R. Talhelm. 1979. Rehabilitating Great Lakes ecosystems. *Tech. Rep.* No. 37, *Great Lakes Fishery Commission,* Ann Arbor, Michigan.

Harris, H.J. and V. Garsow. 1978. Green Bay research workshop proceedings. Univ. *Wisconsin Sea Grant Publ.* WIS-SG-78–234.

Harris, H.J., D.R. Talhelm, J.J. Magnuson, and A.M. Forbes. 1982. Green Bay in the future—a rehabilitative prospectus. *Tech. Rep.* No. 38, *Great Lakes Fishery Commission,* Ann Arbor, Michigan.

Harris, H.J., P.E. Sager, C.J. Yarbrough, and H.J. Day. 1987a. Evolution of water resource management: a Laurentian Great Lakes case study. Intern. *J. Environ. Stud.* 29:53–70.

Harris, H.J., P.E. Sager, S. Richman, V.A. Harris, and C.J. Yarbrough. 1987b. Coupling ecosystem science with management: a Great Lakes perspective from Green Bay, Lake Michigan, USA. *Env. Man.* 11:619–625.

Harris, V.A., and J. Christie. 1987. Nutrient and eutrophication management technical advisory committee report, Lower Green Bay Remedial Action Plan. PUBL-WR-167 87, *Wisconsin Dept. of Natural Resources,* Water Resources Management, Madison, Wisconsin.

Kubiak, T.J., H.J. Harris, L.M. Smith, T.R. Schwartz, D.L. Stalling, J.A. Trick, L. Sileo, D.E. Docherty, and T.C. Erdman. 1989. Microcontaminants and reproductive impairment of the Forster's tern on Green Bay, Lake Michigan—1983. *Arch. Environ. Contam. Toxicol.* 18:706–727.

Richman, S., P.E. Sager, G. Bonta, T.R. Harvey and B.T. Destasio. 1984. Phytoplankton standing stock, size distribution, species composition and productivity along a trophic gradient in Green Bay, Lake Michigan. *Verh. Internat. Verein. Limnol.* 22:460–469.

Sager, P.E., and S. Richman. 1991. Functional interaction of phytoplankton and zooplankton along the trophic gradient in Green Bay, Lake Michigan. *Can. J. Fish. Aquat. Sci.* 48 (In Press).

Smith, P.L., R.A. Ragotzkie, A.W. Andren, and H.J. Harris. 1988. Estuary rehabilitation: the Green Bay story. *Oceanus.* 31(3):12–20.

Chapter 3

Hamilton Harbour Remedial Action Planning

G. Keith Rodgers

> "The Hamilton Harbour RAP is a down-to-earth blueprint for the recla-
> mation of a much abused body of water to allow greater public use and
> enjoyment. It is now on paper! The real challenge will be the transfer
> of this blueprint into reality—no small task which will require major
> political will."
>
> Geraldine Copps
> Alderman
> City of Hamilton, Ontario

Introduction

Hamilton Harbour, one of the original "problem areas" identified
in the 1974 report of the Great Lakes Water Quality Board to the
International Joint Commission (IJC 1974), has a history of environmental problems that spans two centuries and threatens to enter a
third. Growth and prosperity in the region historically have been
linked to the harbor waters and those of nearby Lake Ontario. However, the pre-eminence of three water uses—shipping, industrial
water supply, and municipal and industrial waste reception—imper-
ils the very resource that has sustained the region's economy. Ignorance of environmental consequences, competing water uses and a
jurisdictional labyrinth have combined to thwart an integrated approach to resource management. While pollution abatement has
steadily improved water quality, the lack of a "master plan" acceptable to all users has not been developed, thereby preventing the
complete rehabilitation of the harbor.

The remedial action plan process (IJC 1985; Canada and United
States, 1987) has provided the forum and the sustaining impetus to
develop such a "master plan" to restore and maintain the resource.
This process not only represents an opportunity to implement the
ecosystem approach, called for in the 1978 Great Lakes Water Qual-

ity Agreement (Canada and United States, 1978; Hartig and Vallentyne, 1989), but also provides a test case for the adoption of sustainable development principles in Canada. The purpose of this chapter is to share the experiences of the Hamilton Harbour remedial action plan process.

Background

Physical Geography

Hamilton Harbour is located at the western end of Lake Ontario and is roughly triangular in shape, with an east-west length of eight km (five miles) and a north-south width of five km (3.1 miles) (Figure 8). It has a mean depth of approximately 13 m (42.6 feet), a maximum depth of 24 m (78.7 feet), a volume of 2.8×10^8 m^3 (73,976 million gallons), and a surface area of 22 km^2 (8.5 square miles) (OMOE 1977, 1978). The harbor is connected to Lake Ontario by the Burlington ship canal, which is 820 m (2,690 feet) long, 107 m (351 feet) wide, and 9.5 m (31 feet) deep (Dick and Marsalek, 1973).

Cootes Paradise (occasionally referred to as Dundas Marsh) is an extensive, shallow wetland to the west of Hamilton Harbour composed of 29 percent swamp and 71 percent marsh vegetation (McCullough and Wilson, 1983). It is about four km (2.5 miles) wide and covers an area of 1.2 km^2 (0.5 square miles) (Whillans 1979). It joins hydrologically with the harbor via the Desjardins Canal and is the largest wetland along the western Lake Ontario shoreline (McCullough and Wilson, 1983).

Several streams, including Indian, Grindstone, Spencer, Ancaster, Sulphur and Redhill Creeks, drain a relatively small watershed of approximately 500 km^2 (193 square miles or 22 times the harbor surface area) into the harbor (OMOE 1977). About 58 percent of the area is drained into Cootes Paradise by Spencer, Sulphur and Ancaster Creeks and several smaller tributaries. Natural stream flows and inputs from municipal and industrial sources account for 1.27×10^8 m^{3}y^{-1} (33,553 million gallons per year) and 1.02×10^8 m^{3}y^{-1} (26,948 million gallons per year), respectively (Palmer and Poulton, 1976; OMOE 1978; Harris et al. 1980a), providing an estimated total flow of 80 percent of the harbor volume per year (Palmer and Poul-

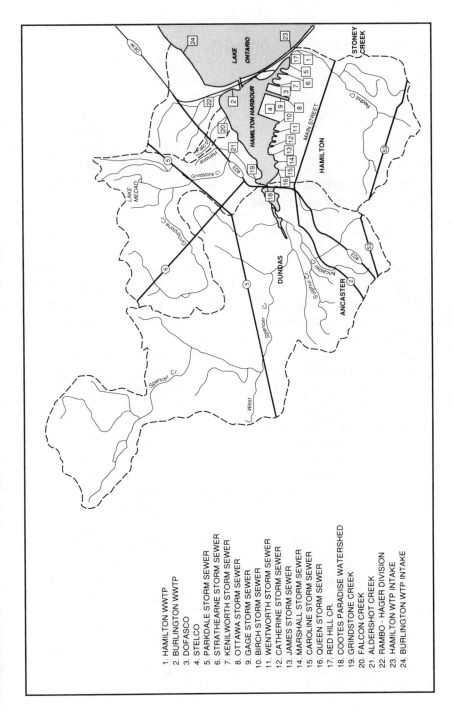

1. HAMILTON WWTP
2. BURLINGTON WWTP
3. DOFASCO
4. STELCO
5. PARKDALE STORM SEWER
6. STRATHEARNE STORM SEWER
7. KENILWORTH STORM SEWER
8. OTTAWA STORM SEWER
9. GAGE STORM SEWER
10. BIRCH STORM SEWER
11. WENTWORTH STORM SEWER
12. CATHERINE STORM SEWER
13. JAMES STORM SEWER
14. MARSHALL STORM SEWER
15. CAROLINE STORM SEWER
16. QUEEN STORM SEWER
17. RED HILL CR.
18. COOTES PARADISE WATERSHED
19. GRINDSTONE CREEK
20. FALCON CREEK
21. ALDERSHOT CREEK
22. RAMBO - HAGER DIVISION
23. HAMILTON WTP INTAKE
24. BURLINGTON WTP INTAKE

Fig. 8. Hamilton Harbour watershed

ton, 1976; Kohli 1979). Untreated storm sewer overflows are estimated to contribute 3.2×10^6 m^{3}y^{-1} (845 million gallons per year) (Ng 1981), while 8.5×10^8 m^{3}y^{-1} (224,570 million gallons per year) are recycled by the industries along the heavily industrialized southern shore (OMOE 1978) (Figure 8).

Water exchange with Lake Ontario keeps the harbor balanced hydraulically. Kohli (1979) calculated that an average of 2.04×10^6 m^{3}d^{-1} (539 million gallons per day) of harbor water flowed into Lake Ontario and 0.73×10^6 m^{3}d^{-1} (193 million gallons per day) flowed upstream, accounting for a net exchange rate of 0.5 percent of the harbor volume per day into the lake.

Political Geography and Jurisdiction

Two cities—Burlington on the north and Hamilton on the south—are immediately adjacent to the harbor. Most of the watershed is contained, however, within the counties of Halton and Wentworth, which includes four major towns (Ancaster, Dundas, Stoney Creek and Waterdown) and several smaller ones.

Several city and town councils, two regional governments, and provincial and federal governments exert direct or indirect control on development and growth within the basin and management of its renewable and nonrenewable resources. Therefore, a complex maze of overlapping responsibilities for the harbor's water quality and environmental protection exists among numerous jurisdictions and agencies. In addition, a separate Hamilton Harbour Commission controls shipping and navigation. The Commission includes two federal representatives and one member from the City of Hamilton, and was created by an Act of Parliament in 1912. Cootes Paradise, once a Crown Game Preserve and the major public access to the waterfront, is privately owned by the Royal Botanical Gardens. A detailed account of the political jurisdictions and the various environmental objectives and guidelines affecting water quality in the harbor is included in the Stage 1 RAP for Hamilton Harbour (Canada-Ontario Agreement 1989).

The various levels of government, combined with overlapping mandates, responsibilities and ownership, have effectively blocked integrated environmental or ecosystem planning in the past and have created some adversarial problems.

History

Humans have inhabited the watershed of Hamilton Harbour for more than 3,000 years (Scott 1970). The first white man, LaSalle, did not "discover" the harbor until 1669 and for over a century the harbor saw few if any white men. In the late 1700s European settlers moved into the area and cleared the land. Gradually the watershed was transformed: the forest was cut down, the land was tilled, streams were dammed and their power harnessed, and mills were built.

To accommodate the transport of supplies for and produced by the growing population, the navigational approaches to the harbor and to the Town of Dundas were improved. By 1827, a new shipping canal pierced the sandspit separating the harbor and Lake Ontario, constructed in the location of the present canal (Figure 8). By 1837 the Desjardins Canal, which provides passage through the western heights separating Cootes Paradise from the harbor, was completed. The marsh, harbor and lake were now directly linked and all natural refuges laid bare to exploitation. Iron foundries were established in Hamilton and a steady industrial growth ensued, and was accelerated with the completion of the Great Western Railway through the city in 1854.

Throughout this period the population base grew rapidly. This growth, combined with poor sanitary facilities in the growing city, resulted in two major cholera epidemics in 1832 and 1854. The city responded by constructing its first sewers, which emptied directly into the harbor and nearby municipal drinking water supply from Lake Ontario.

Due to the industrial growth and shipping facilities at the water's edge, surrounding wetlands and the long shallow inlets along the harbor's south shore were filled and much of the natural fish and wildlife habitat was lost. Filling continued into this century, often at rapid rates. Between 1926 and 1959, an estimated 22 percent of the open water area of the harbor was lost to filling (Cairns 1985).

Pollution from the urban and industrial growth had noticeable effects on a thriving commercial and sport fishery (Holmes 1986). Overfishing, habitat loss, water pollution and economic misfortunes all contributed to the virtual collapse of the commercial fishery in the early 1900s and the sport fishery by 1940 (Holmes 1986).

Throughout the first 60 years of the 20th century, the deteriorating condition of the harbor was a relatively unimportant issue to area residents and regional governments. Two world wars, a major depression and several economic recessions, followed by a sustained post World War II growth boom, took precedence.

Water Quality and Pollution Control

Beginning in the late 1950s, several studies on water quality and biological effects were conducted by the Ontario Water Resources Commission, precursor to the Ontario Ministry of the Environment. These studies and the necessary legislative powers provided by the Ontario Water Resources Act were used to enact municipal and industrial source control and water quality improvements. A primary sewage treatment facility was constructed in 1963 and secondary treatment was introduced in 1972. Despite these changes, growth in the area and concomitant increases to pollution loads brought continued water quality deterioration until the mid-1970s.

More detailed studies on the harbor, Cootes Paradise and the various point and nonpoint sources of nutrients and toxic substances were undertaken by the Ontario Ministry of the Environment from 1970 to 1985. Investigations monitored water and sediment quality, the causes of oxygen depletion in bottom waters, the relationships among algal biomass, bacteria, nutrients and abiotic physical factors, and the sources and fate of toxic contaminants. These studies provided a clearer picture of the complex relationships among factors influencing water quality; however, they largely ignored other ecosystem factors such as habitat and health of the aquatic community. During the same time, approximately 260 million dollars were spent by industry and the municipalities to abate water pollution and, as a result, water quality slowly improved. However, several problems still were evident, and it became obvious the piece-meal approach to pollution abatement was inadequate. A master plan that could be used to guide further development or changes in the watershed was essential.

RAP Development

In September 1985 the Ontario Ministry of the Environment released a report summarizing the findings of its extensive study program, indicating what further information was required, and outlining some general management options (Ontario MOE 1985). The report concluded that while water quality had improved, it remained severely degraded. Extensive improvements would be necessary to bring water quality to Ministry and Great Lakes Water Quality Agreement standards. These improvements would be costly and thus factors such as the present significance, future potential and desired uses of the harbor must be considered in choosing remedial options. An "implementation" or review committee was convened to examine the issues and draw up a plan for the harbor's future. Within six months of the release of the Ministry's report and announcing their intentions to "get the interested parties together," Ontario adopted the IJC's remedial action plan (RAP) process for Hamilton Harbour.

In June 1986, a facilitator was contracted to survey interest in and commitment to an advisory committee called the Stakeholders. Membership was drawn from local elected officials, regional, provincial and federal agency representatives, the Hamilton Harbour Commission, the Royal Botanical Gardens, environmental groups (local and provincial), citizen groups, representatives of the major industries, academics, union representatives and special interest groups with a stake in the harbor's water quality.

The 46 stakeholders met at a two-day workshop, followed by a public meeting, in July 1986. The stakeholders' ideas and intent were compiled in an Interim Report of the Stakeholders in September 1986 (Hamilton Harbour Stakeholders 1986), in which the group stipulated the principles, special issues, water use goals and concepts for RAP development and implementation (see Table 4). They also declared their interest in continuing as advisors during RAP development, although no one probably realized the likely duration of that commitment. Fifty-nine percent of the original membership has stayed with the process. Total membership, after almost four years, stands at 43, with 38 of the original groups or agencies represented. Five new groups or individual members have been added, and the stakeholders operate with control over their own membership, a five-person executive (but no single chairperson) and six committees (ecosystem, socio-economic, technical options, implementation-institutions, membership, and communications).

A technical team of nine members oversees the technical study program, provides advice on proposed implementation measures, and prepares the RAP at its various stages and for the public consultation process. This team is appointed by the RAP Steering Committee of the Canada-Ontario Agreement (COA) Review Board which, in turn, is charged with managing the overall RAP program to which Canada committed under the revised 1978 Great Lakes Water Quality Agreement. This team has been greatly aided by colleagues in their respective agencies, and by staff of several of the organizations represented on the stakeholder group.

TABLE 4.   Stakeholders' Principles, Issues and Goals for Hamilton Harbour RAP Development

---

1. Water Quality—Approaches and General Principles
   1.1 The Principle of an Ecosystem Approach
   1.2 The Principle of Human Health
   1.3 The Principle of Public Acceptance and Support for Remedial Actions
   1.4 The Principle of Access
   1.5 The Principle of Aesthetics

2. Considerations Relating to Water Quality Enhancement
   2.1 Harbor Water: Impacts on Lake Ontario
   2.2 The Windermere Basin

3. Restricted Uses and Actions
   3.1 Shoreline Filling
   3.2 Waste Water Receiving Body

4. Current and Preferred Water Use Goals and Remedial Actions to Achieve Goals
   4.1 Introduction
   4.2 Defined Use Goals
   4.3 Specific Goals and Actions to Permit Defined Uses
        4.3.1 Recreational Boating
        4.3.2 Water Sports
        4.3.3 Shipping and Navigation
        4.3.4 Industrial Uses
        4.3.5 Waste Water Receiving Body
   4.4 Enhanced and Future Use Goals and Remedial Actions to Achieve Goals
        4.4.1 Fisheries
        4.4.2 Wildlife Appreciation and Habitat Appreciation
        4.4.3 Swimming
        4.4.4 Educational Resource

5. Issues Related to Plan Development and Implementation Council for the Harbor

6. In 1987 the principle of zero discharge/virtual elimination for persistent toxic chemicals was linked with the ecosystem approach principle as the two primary principles to be applied in the RAP.

The technical team presented an interim report to stakeholders in March 1987 outlining tentative costs and identifying crucial information gaps that affected its ability to predict the response of the harbor to various contaminant controls (Canada-Ontario Agreement 1987). Soon thereafter, guidance on a RAP's content and beneficial use criteria were provided by the IJC's Great Lakes Water Quality Board (WQB) and the COA Review Board. This guidance resulted in an expanded technical description of the environmental problems that included more biological information, especially on the fisheries, and expanded data collection and research to address the information gaps identified by Ontario MOE (1985). Key among these studies were assessments of contaminated sediments and efforts to develop a better understanding of the link between contaminants and biota and among trophic conditions and the loading of suspended solids, ammonia, phosphorus and direct oxygen-demanding substances.

A discussion document was released in March 1988 describing environmental conditions, the goals of the program and the broad range of technical options (Canada-Ontario Agreement 1988). Considerable written advice was provided by the stakeholders, a further public meeting was held and stakeholder subcommittees were formed to explore key issues.

Based on this advice, the Stage 1 RAP on environmental conditions and problem definition was completed in March 1989 and submitted for COA review and subsequent IJC commentary (Canada-Ontario Agreement 1989). Parallel with the development of the Stage 1 RAP, stakeholders and the writing team explored a series of recommendations for technical remedial options, basinwide planning options, institutional arrangements during the implementation phase and educational concerns. This consultation was completed and led to a draft RAP for broader public consultation in late 1990, before its finalization and melding with agency commitments in 1991.

The remedial action planning process is being carried out concurrent with ongoing pollution abatement actions (Hartig and Zarull, 1990). The two major iron and steel industries are in the intensive monitoring phase of the province's Municipal-Industrial Strategy for Abatement (MISA) program, while the discharge of toxic contaminants to the municipal sewer system by business and industry is being inventoried and assessed under another component of this program. One contaminated section of sediment in the harbor, Windermere Basin, has been dredged and confined. Pollution control planning studies are moving ahead for the City of Hamilton and a program for remediation of combined sewer overflows has already addressed 25 percent of the overflow volume that reaches the harbor.

In addition to these remedial actions, other developments could significantly impinge on the RAP. Marinas, parks, highway construction and marsh restoration projects could have a direct impact on the shoreline and embayments of the harbor.

The technical team and stakeholders have intentionally obtained updates and briefing from project proponents of these developments. The integrated and comprehensive nature of the RAP has generated public interest and concern that current developments not preclude improvements that might be expected by the RAP, and interest in harbor conditions has given rise to a better appreciation of the broader consequences of various types of proposed development.

Pilot scale and full-scale experimentation with additional sewage treatment technologies have also taken place as a result of reports outlining the relationship between trophic conditions and provisional nutrient loading targets. Nutrient loadings to the harbor have been reduced, which in turn has made it possible to verify the trophic response of the harbor to incremental loading reductions. Thus, there is greater confidence in specifying the required final loading targets.

The stakeholder advisory committee, with whom the technical team has worked, represents the many agencies and political jurisdictions involved with the harbor, as well as other citizens. Many interests have been brought together that would not otherwise collaborate on matters of such scale. Since the RAP is comprehensive, the RAP has become a "lightning rod" to discussing old and new issues regarding the harbor. While not all these issues have been resolved, the airing of positions and a focus on the future of the harbor ensures a reasonable degree of common cause.

The RAP process has been an educational experience for agencies and citizen members, and particularly for the science specialists and those who see the harbor in more aesthetic terms. A translation service occasionally is needed to facilitate communication between all interested parties.

Some stakeholders entered the process with high expectations to influence the consultative process. While these expectations may still be realized, the time it has taken to work through the process has dampened some spirits. Detailed consultations and completing essential studies to adequately define the problem and evaluate remedial options takes time, and other undeveloped or unofficial standards that have been added for the RAP reports are delaying matters as well.

Key to many concerns will be the final answers given by authorities for questions about how the plan will be formalized or legalized and how the remedial work will be financed. Those negotiations and discussions are underway.

Gaps in scientific information that assess effective options to deal with contaminated sediments also have delayed the Hamilton Harbour RAP (and probably most other RAP teams as well). Principle questions concern which contaminants have the worst impact, what are the food chain effects, which removal techniques have the least environmental impact, and what are the suitable treatment and disposal technologies. Harbor sediment data from bulk chemical analyses, bioassays and cores (i.e. subsurface) have prompted a demand for a decision on cleaning up at least one area where there are unusually high concentrations of polynuclear aromatic hydrocarbons. It promises to be one test case for this kind of decisionmaking for Canadian Areas of Concern.

Hamilton Harbour has only one outlet to Lake Ontario via the Burlington Ship Canal (Figure 8). All watershed streams, effluent from the steel industries, effluent from sewage treatment plants, urban runoff and groundwater seepage drain into the harbor. No effluent from the watershed is discharged directly to Lake Ontario—all of it passes through the harbor—and this has placed tremendous stress on the harbor. In combination with infilling of 25 percent of the area of the harbor since 1909 (particularly marsh and shallows habitat), these stresses have radically affected the aquatic biota. Habitat management and rehabilitative measures that might be taken for biota have not been tested or are not, at least, considered to be standard procedures.

As so often is the case in environmental restoration work, predicting the benefit of some measures is difficult. Where those uncertainties exist, a staged approach has been recommended and monitoring is designed to assess the outcome impact of each stage of the plan.

Finally, one must look to the future and the ability to gauge the effort necessary to maintain the balance of social, economic and environmental concerns in a sustainable way. General consensus is that restoration of the system to some reasonable condition over the next 10 to 20 years is within grasp; however, if the impact of contaminated sediment is more pernicious or less reparable than currently believed, a more lengthy time of recovery may be needed. Efforts to correct source discharges, notwithstanding the best application of reduction, reuse and recycling that is achievable, does involve the use of more chemicals and more energy. This would suggest another, long-term strategy related not only to the goods and services produced and consumed, but also to marked changes in societal lifestyles.

The long-term prospects for the harbor are less amenable to quantitative analysis. Given contaminant loading targets commensurate with a rehabilitated environmental state for the harbor, or for the western end of Lake Ontario, the technology available to compensate for population growth or for a combination of population growth and traditional economic development is foreseen to be too expensive per capita unless communities choose to channel their development in quite different ways. Strategies have been developed to help communities find different ways to achieve their desired future state and minimize the potential for additional unpleasant surprises.

Conclusions

While the process of developing the RAP has not been as quick as originally envisaged, it has provided an opportunity for mutual education among many sectors of the region's community and has developed the information required to aid in community decision-making. In order to finalize the RAP, some exact solutions to certain problems will be missing, or the definition of some problems will not yet be complete. It will be essential to pursue investigations almost continuously if further surprises are to be avoided.

It is uncertain what the consequences will be for situations where community objectives do not meet as yet undefined criteria for defining "How Clean is Clean?" Technically speaking, no RAP is valid until those criteria are established. These criteria are being established by different constituencies, and while some conditions are clearly of broader regional concern (i.e. the discharge of persistent toxic chemicals), other localized conditions under discussion are not. The consequences for communities facing extraordinary costs that flow from benefits to a much larger community seem uncertain. A rigorous application of the user-pay principle, for example, quickly reveals tentacles that reach far beyond the local community.

The RAP process to date suggests strongly that it must continue in a formal way for some time. The Stage 1 and Stage 2 RAP reports required under the revised 1978 Great Lakes Water Quality Agreement will always be incomplete. Eternal vigilance, continued action and public consultation will be required to sustain the vision embodied in that Agreement.

REFERENCES

Cairns, V. 1985. *Hamilton Harbour: Status and Rehabilitation Potential.* Presentation to the Great Lakes Science Advisory Board. September 13, 1985. Burlington, Ontario, Canada.

Canada-Ontario Agreement (COA). 1987. *Interim Report of the Writing Team for the Hamilton Harbour Remedial Action Plan.* Burlington, Ontario, Canada.

Canada-Ontario Agreement (COA). 1988. Goals, Problems and Options—A *Discussion Paper for the Hamilton Harbour Remedial Action Plan.* Burlington, Ontario, Canada.

Canada-Ontario Agreement (COA). 1989. *Stage 1 Report, Environmental Conditions and Problem Definitions for the Hamilton Harbour Remedial Action Plan.* Burlington, Ontario, Canada.

Canada and United States. 1978. *Great Lakes Water Quality Agreement of 1978.* Windsor, Ontario, Canada.

Canada and United States. 1987. *Protocol to the 1978 Great Lakes Water Quality Agreement.* Windsor, Ontario, Canada.

Dick, T.M., and J. Marsalek. 1973. Exchange Flow Between Lake Ontario and Hamilton Harbour. *Environ. Can., Inland Wat. Dir., Sci. Ser.* No. 36. Burlington, Ontario, Canada.

Hamilton Harbour Stakeholders (HHS). 1986. *Interim Report—Hamilton Harbour's Water Quality: The Stakeholder's Proposals.* Prepared by Land Use Research Associates. Burlington, Ontario, Canada.

Harris, G.P., B.B. Piccinin, G.D. Haffner, W.J. Snodgras and J. Polak. 1980. Physical variability and phytoplankton communities: I. The descriptive limnology of Hamilton Harbour. *Arch. Hydrobiol.* 88:303–327.

Hartig, J.H. and J.R. Vallentyne. 1989. Use of an ecosystem approach to restore degraded areas of the Great Lakes. *AMBIO.* 18:423–428.

Hartig, J.H. and M.A. Zarull. 1991. Methods of restoring degraded areas of the Great Lakes. *Rev. Envir. Cont. Toxicol.* 117:127–154.

Holmes, J.A. 1986. *Rehabilitation of the fishery in the Hamilton Harbour–Cootes Paradise Ecosystem.* M.Sc. Thesis, Dept. Zoology and the Institute for Environmental Studies, University of Toronto, Toronto, Ontario, Canada.

International Joint Commission. 1974. *Report of the Great Lakes Water Quality Board.* Windsor, Ontario, Canada.

International Joint Commission. 1985. *Report of the Great Lakes Water Quality Board.* Windsor, Ontario, Canada.

Kohli, B. 1979. Mass exchange between Hamilton Harbour and Lake Ontario. *J. Great Lakes Res.* 5:36–44.

McCullough, G. and J. Wilson. 1983. Wetlands evaluation of Cootes Paradise wetland. Unpub. rep. compiled June 13, 1981. *Canadian Wildlife Service.* Burlington, Ontario, Canada.

Ng, P.S. 1981. *Models for the hypolimnetic oxygen deficit in Hamilton Harbour.* M. Eng. thesis, Dept. Chem. Eng., McMaster University, Hamilton, Ontario, Canada.

Ontario Ministry of the Environment (OMOE). 1975. *Hamilton Harbour Study 1975.* Ontario Ministry of the Environment, Water Resources Branch, Toronto, Ontario, Canada.

Ontario Ministry of the Environment (OMOE). 1978. *Hamilton Harbour Study 1976.* Ontario Ministry of the Environment, Water Resources Branch, Toronto, Ontario, Canada.

Ontario Ministry of the Environment (OMOE). 1985. *Hamilton Harbour Technical Summary and General Management Options.* Ontario Ministry of the Environment, Water Resources Branch, Toronto, Ontario, Canada.

Palmer, M.D. and D.J. Poulton. 1976. Hamilton Harbour: productivity of the physiochemical processes. *Limnol. Oceanogr.* 21:113–127.

Threader, R.W., P.O. Hodson and V.W. Cairns. 1985. *The history of Hamilton Harbour and its limnology.* Unpub. report. Department of Fisheries and Oceans. Burlington, Ontario, Canada.

Whillans, T.H. 1979. Historic transformations of fish communities in three Great Lakes bays. *J. Great Lakes Res.* 5:195–215.

Chapter 4

Restoring the Rouge

Roy Schrameck, Margaret Fields, and
Margaret Synk

"RAPs are the building blocks of the international effort to protect and
restore the Great Lakes. They are critical to revitalizing the Great Lakes'
most polluted areas. The Rouge River RAP is an excellent example of
how local citizens can not only participate, but accelerate the cleanup
of these sites. The RAP's grassroots involvement has revived agency
enthusiasm, promoted creative initiatives, and energized local officials,
businesses and residents. Its education programs for school children
have become a model throughout the region. The Rouge River RAP
shows us how clean water goals can not only help renew our environment, but our communities and local economy as well."

Carl Levin
United States Senator

Introduction

The Rouge River flows through metropolitan Detroit, Michigan and
empties into the Detroit River. Like many other Great Lakes tributaries, it was used by early settlers as a source of drinking water and
recreation and a means of transportation for fur trade and supplies.
Its location encouraged a rapid population growth and industrialization, the same actions that ultimately became its downfall.

The Rouge includes four branches totalling 202 km (126 miles)
winding through 47 communities, three counties, and over 80 km
(50 miles) of public parkland in southeast Michigan (Figure 9). The
watershed covers a 1,210 km² (467 square mile) area that is home
to 1.5 million people and contains 404 lakes and ponds.

Unlike other rivers in Michigan, the Rouge winds through the
state's most populated and industrialized area. It is not surprising,
then, that the river is considered one of the most polluted rivers in
Michigan.

In spite of its degraded conditions the Rouge still attracts substantial recreational use, particularly at the Middle Rouge Parkway,
one of the first parkway systems in the United States. This 28 km

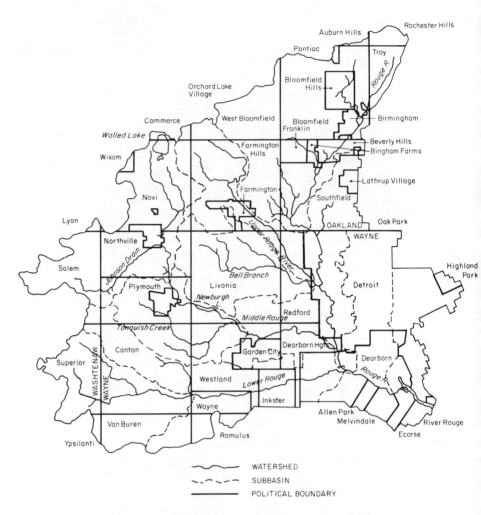

Fig. 9. The Rouge River basin, southeastern Michigan

(17.5 mile) stretch of the river starts in Dearborn, near the mouth of the Middle Branch, and extends into Northville in the headwaters. The Middle Branch has four lakes developed by Henry Ford to provide power for the village industries built on each lake. Today the industries are gone and the area is a popular parkway that supports biking, jogging, walking, softball, volleyball, picnicking and other recreational activities. While there are fishing holes, some fishing and all body contact water activities, such as swimming, are discouraged. A summary of impaired uses in the Rouge River by subbasin is presented in Table 5.

In August 1986, the Michigan Department of Public Health issued a fish consumption advisory for bottom feeding fishes due to PCB contamination. This advisory applied to the lower 8 km (five miles) of the Rouge River, which is considered the "industrialized" portion of the basin. By 1989, further fish contaminant surveys resulted in the consumption advisory being extended upstream into the residential areas of the Middle and Lower Rouge for bottom feeding fish, such as carp and sucker, and some sportfish on the Middle Rouge. Recent Michigan Department of Natural Resources (DNR) sampling has found no significant active point sources of PCBs; rather, sediment contaminated from historical discharges is the primary source of PCB contamination of the fishery (Evans 1990).

Contaminated sediments from historical discharges still plague the Rouge River, but some problems are also due to ongoing discharges. Due to the unpredictability and frequency of the many combined sewer overflow (CSO) discharges along the Rouge River, the Wayne County Health Department has issued a standing health advisory against total body contact for all portions within its boundaries. The pollution and bacterial contamination issues facing the river were perhaps best summarized in a 1986 news article headline which read, "People found the Rouge, liked it, settled it and slowly started killing it." In fact, some people advocate enclosing the Rouge as a sewer. A proposal developed by Wayne County Parks and funded by the State of Michigan has shown, however, what the Rouge could be through remedial action and restoration (Wayne County 1990).

The county study outlined the full potential of the Middle Rouge Parkway and found that wildlife habitat could be improved, native wild flowers could be re-established, fish could be stocked to make the parkway more enjoyable for anglers, natural resource education could be encouraged and recreation expanded, including canoeing along the 5 km (three mile) stretch from Wilcox Lake to Newburg Lake. These benefits could all be accomplished and maintenance costs would decline as a result.

Combining optimum recreational use while also protecting wildlife habitat has become a community vision for the Rouge River. This chapter highlights the remedial action plan process being used to restore the Rouge River to accomplish the citizens' vision for this waterway.

Early Restoration Efforts

The river's name came from the French, because of its natural red clay color. By the 1960s, however, the Rouge was flowing orange due to the discharge of large quantities of iron in industrial pickle liquor wastes (Cowels 1975). The orange color was evident when a boat cut a wake through the heavy black waste oil floating on the surface.

Such problems were clearly due to industrial discharges, which were, at least visibly, having the worst impact on the surface waters. Restoring the Rouge thus began with controls for industrial

TABLE 5.   Summary of Impaired Uses for the Rouge River

| | Designated Uses | | | | |
Subbasin	Water Contact	Water Contact	Warm Water Fishery	Canoeing Navigation	General Aesthetic
Main 1 Southeast Oakland Co.	●	⊙	●	●	●
Main 2 Southfield/ Detroit	●	⊙	●	●	●
Main 3 Detroit/ Dearborn	●	●	●	●	●
Main 4 Detroit/River Mouth	●	●	⊙	○	●
Upper 1 Farmington Hills/Farmington	●	●	●	⊙	●
Upper 2 Livonia/Bell Branch	●	●	●	⊕	⊕
Middle 1 Novi/ Northville	●	●	●	○	●
Middle 2 Plymouth/ Westland	●	●	●	●	●
Middle 3 Garden City/ Dearborn Heights	●	●	●	●	●
Lower 1 Superior Twp./ Canton Twp.	●	●	●	●	●
Lower 2 Wayne/Inkster	●	●	●	●	●

○ no impairment   ⊕ insufficient data   ⊙ occasional or slight impairment   ● frequent or severe impairment discharges. Michigan initiated an industrial control program in the early 1960s and issued Orders of Determination and Stipulations to industries that required pollution abatement. In the early '70s, the state joined the federal government's newly developed National Pollutant Discharge Elimination System Permit Program (NPDES). These joint state/federal permits replaced the earlier orders and required more extensive abatement programs. By the 1980s, industries were no longer a major source of pollutants to the Rouge. The river no longer flowed orange, and floating oils no longer caught fire as they did during the 1960s.

Although many thought that industrial controls would result in a clean river, the Rouge River was still Michigan's most polluted waterway. Much of the river still did not meet the state's water quality standards for municipal discharges (primarily combined sewer overflows), which became visible as industrial pollution lessened (Jackson 1975). Historically, sewers were built to protect human health and safety, not the environment. The first sewers were designed to direct disease-carrying sanitary wastes away from populated areas and to reduce flooding from storms. Known as combined sewers, these pipes discharged directly into the closest river. The long-term environmental impact of this action was not recognized when they were first built.

Once the impact on drinking water and recreational uses was recognized, interceptor sewers were installed to collect the dry weather flow of water and carry it to a wastewater treatment facility. However, since a finite system cannot transport an infinite quantity of flow, the system discharged excess flow into the river during storms; otherwise the flow would back up into homes during hard rainfalls. As the population increased, the frequency of these discharges naturally increased as well. As greater urbanization of the region resulted in more paved surfaces, stormwater runoff also increased. By the 1980s, some systems were discharging to the river during little or no rainfall.

During the 1970s, Michigan DNR began to require that all new sewer systems be installed separately: one for sanitary wastes and one for stormwater. Even with this provision, approximately 168 combined sewer overflows (CSOs) discharge into the river. Some overflow from separated sanitary sewers still occurs when the combined sewer systems surcharge and cause the connecting separated systems to back up. These CSOs may also be discharging industrial wastes from industries within the tributary collection area.

The legacy of industrial discharges, CSOs and sanitary overflows is sediments contaminated with heavy metals and PCBs (Hartig 1984). Several sportfish in the Middle Rouge basin are listed in the "no consumption" advisory category due to contamination by these substances. Increased development and runoff have resulted in widely fluctuating water levels that cause erosion, log jams and sedimentation. Sanitary and CSO discharges have caused fish kills and there is a standing advisory against any type of body contact with Rouge waters. "Sludge beds" below CSOs cannot support a suitable aquatic environment, and floating mats of sludge move downstream from major CSOs in the summer months. Bubbles of hydrogen sulfide from decomposing organic bottom sludges escape to the surface of the river and produce a "rotten egg" odor. These conditions demanded that the State of Michigan initiate a correction program that would reverse the decay of the Rouge River and restore it to a useable natural resource.

The Rouge Strategy

Isolated corrective efforts were documented as early as 1973. Michigan DNR and community agencies, such as the Southeast Michigan Council of Governments (SEMCOG), tried to assess the river's condition and compiled reports and potential plans. Scattered individual communities organized cleanup efforts on their river segments. Michigan DNR and the United States Environmental Protection Agency (U.S. EPA) also performed studies on the Rouge River, but these independent efforts were not coordinated with one another.

In the early 1980s several complaints were filed concerning CSO discharges. At the Hubbell-Southfield overflow in particular, Michigan DNR coordinated with one county and two local agencies to correct problems at this discharge location. Soon thereafter, the agency recommended that a basinwide approach be developed to correct the Rouge River's CSO problems. By the end of 1984, Michigan DNR and U.S. EPA staff began coordinating studies and starting work on the Rouge River Strategy. These were the beginnings of a coordinated, basinwide effort that was sound technically and supported politically.

The character and complexity of the Rouge River problems and potential solutions were recognized early in the process. Clearly, no one community's action would seriously improve conditions if numerous CSO discharges were upstream; no one agency could act unilaterally and effectively to correct the multifaceted Rouge problems. Only a coordinated effort by all communities and agencies would have the desired impact.

It was also recognized that the river could be studied ad infinitum and yet not lead to any real improvement. A Rouge River Basin Strategy was developed to initiate a coordinated process where necessary studies and remedial projects could be implemented concurrently. It was this strategy that led to the development of a remedial action plan (RAP).

On October 1, 1985, the Michigan Water Resources Commission (the state agency responsible for protecting the use and quality of surface, ground and Great Lakes waters by establishing rules and issuing permits or enforcement orders) passed a resolution implementing the Rouge River Basin Strategy, action plan and public participation process to abate water pollution in the Rouge River. The commission instructed Michigan DNR staff to develop a RAP and implement a public participation process for local communities in coordination with their strategy.

The Rouge River Action and Remedial Action Plans

This Rouge River Action Plan focused on actions needed immediately, including:

- designating the Surface Water Quality Division of Michigan DNR as the lead agency to organize a large, coordinated agency/community effort;
- establishing the Rouge River Executive Steering Committee, Rouge River Basin Committee and Technical Committees to facilitate this effort;
- appointing a full-time Rouge River Coordinator within Michigan DNR;
- assigning Michigan DNR's Detroit District office as technical staff support;

- contracting with Southeast Michigan Council of Governments (SEMCOG) to assist in coordination and development of the RAP;
- establishing long-term goals;
- identifying and assessing available data and determining future monitoring needs;
- identifying known pollution sources and alternative solutions;
- evaluating all permitted dischargers within the basin;
- expanding the isolated community Rouge River cleanup campaigns to a basinwide program; and
- implementing known and necessary corrective actions immediately.

The public participation process initiated by the action plan was unique to the RAP process at the time. The process recognized that, while the state had the necessary clout to force corrective programs for the Rouge basin, the enforcement approach would not establish local ownership of the RAP and could result in lengthy litigation that actually delayed river cleanup. Many local participants in the Rouge cleanup had equal legislative and political influence as did Michigan DNR, which in combination could accomplish much more than separate programs could ever provide.

To facilitate public participation, two local committees were formed (i.e. the Rouge River Basin Committee and the Executive Steering Committee). The Rouge River Basin Committee provides a means to inform and obtain input from local governments and other local interests within the basin. It serves as a local consensus-building organization and as an advisor to the Executive Steering Committee and Michigan DNR, which has the overall responsibility to develop and implement the Rouge River RAP.

The basin committee's members come from all communities within the Rouge basin, from the legislative and executive branches of state government, and from the Michigan Natural Resources Commission and the Michigan Water Resources Commission. Representatives from the Rouge River Watershed Council, the International Joint Commission, and public and industrial interests such as the League of Women Voters, Michigan United Conservation Clubs, East Michigan Environmental Action Council, Michigan

Clean Water Coalition and the Michigan Manufacturers Association also serve on the committee.

The Executive Steering Committee includes the key decision-makers within the basin (representatives from the Water Resources Commission, the Governor's office, DNR, U.S. Environmental Protection Agency, SEMCOG, the City of Detroit, the counties of Wayne and Oakland, the Rouge River Watershed Council, four local government representatives and a local citizen representative). It coordinates project implementation and provides direction and guidance to the planning effort. It also serves in an advisory role to Michigan DNR and as a direct advisor to the Michigan Water Resources Commission.

Committee efforts in planning and reaching agreement on appropriate remedial actions for the RAP began in earnest in 1988. A first draft was completed for review in May 1988 (SEMCOG 1988) and later approved by the committees in January 1989. The Michigan Water Resources Commission received the RAP for review in February 1989, which adopted the plan as the guide to restore the river. The Rouge River RAP truly represents a systematic and comprehensive ecosystem approach to identifying problems and solutions for the entire Rouge River watershed. A timetable of recommended actions is presented in Figure 10.

In addition to establishing the framework for developing and implementing the Rouge River RAP, the strategy established an operational premise: those projects that would contribute to cleaning up the Rouge and were in a position to move ahead should not be delayed by additional studies, committee meetings or the RAP planning process. This was an innovative approach for ongoing project implementation, but was consistent with the generally aggressive "move-ahead" attitude expressed by the Michigan DNR and the Basin and Executive Steering Committees.

Current Restorative Efforts

Rouge Rescue and Community Efforts

Several independent yet complementary and cooperative community and citizen efforts are underway. Friends of the Rouge was

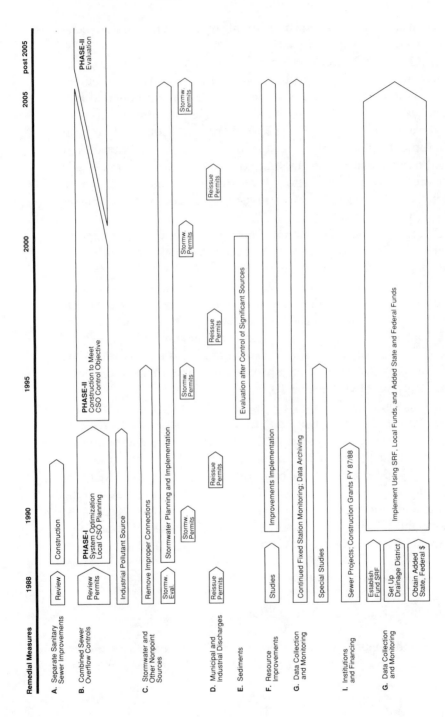

Fig. 10. Summary timeline of Rouge RAP implementation

formed as a nonprofit organization in 1986 to organize a basinwide Rouge River Cleanup Day and promote public education and involvement in cleanup. Prior to 1985, some communities such as Southfield, cleaned up their portions of the river in isolated efforts. The June 7, 1986 "Rouge Rescue," as the cleanup days are called, was the first cooperatively organized cleanup among all Rouge communities. "Rouge Rescue" days have been organized and held on the first Saturday in June every year since 1986.

Organizing *Rouge River Days* is no small feat. The 1989 rescue involved 20 municipalities and several Wayne County locations that required site management. Because these cleanups are in a polluted waterway and there is a potential for exposure to CSO discharges while trying to remove log jam materials, only government or contractor employees are allowed in the river. Refreshments, first aid, trash bags, chain saws, cranes, dumpsters, chippers, landfill disposal, municipal personnel and other supplies must be organized. The costs for organization, promotion, equipment and disposal could be substantial, but so far have been minimal due to volunteers and donated monies and services.

The level of volunteer help has been heartwarming. In 1989, 2,700 citizens devoted their Saturday to help the Rouge, and donations have been substantial as well. The Ford Motor Company donated $30,000 and WJBK-Television donated all air time to help get the rescue days started in 1986. Since then the television station has continued its promotional efforts to help make it a visible and successful effort. Proceeds from an annual basketball game between WJBK-TV and the Detroit Water and Sewerage Department are also donated to the Friends of the Rouge and the Rouge Rescue. Other private services or materials have been donated by companies such as Hygrade, Faygo, Michigan Tractor & Machinery, United Parcel Service, Waste Management, Inc., City Environmental, and others.

What has the Rouge Rescue accomplished? In 1989, the following was removed from the river: 2,905 m³ (3,800 cubic yards) of debris, 78 logjams, four cars, several miscellaneous car parts, a rug, 10.7 m (35 feet) of steel pipe, a washing machine, three televisions, one coca-cola dispensing machine, three picnic tables, an old shed in pieces, an exercise bike, two 1.8 m (6 foot) diameter tractor tires, many shopping carts, two 209 L (55 gallon) drums, and several styrofoam chunks. This is fairly typical of the quantity and types of materials collected in the three earlier Rouge Rescues. The events are important because they remove unsightly trash, open the river for better flow and improve the potential for fisheries and recreation, and provide first-hand experiences for people on what pollution they have caused. It is not only educational, but also encourages people to take pride in their river.

Public Education and Monitoring

In addition to the Rouge Rescue, Friends of the Rouge is active in public education. One project is coordination and funding of the Rouge River Interactive Water Monitoring Project organized by Friends of the Rouge in cooperation with the University of Michigan for high school students. The University of Michigan provides mainframe computer support and a large team of support staff for the project, which involves 52 science classes monitoring nine water quality parameters in the Rouge River. The students share their data by computer and a Water Quality Congress is organized annually to bring students from all 52 classes together to interpret data, learn from each other's experiences and develop proposals for action. It is hoped that this program will expand to over 100 schools within the next few years.

Michigan's ambient monitoring program is similar to the multi-station sampling effort in the high schools. Like the Rouge Rescue, private donations helped with the program success. The ambient monitoring program involved weekly intensive sampling and analysis over a two-year period. Laboratory support was provided by Michigan DNR, the Detroit Water and Sewerage Department and Detroit Edison.

The State of Michigan has also funded monitoring of the Rouge River under its Public Act 307 program, which designated much of the Rouge River as an Act 307 site. As a result of that designation, funds have been provided to study and correct the contaminated sediment problem and identify sources. As a result, Michigan DNR has received funds for monitoring all industrial dischargers and performing fish surveys.

One fish study recently examined the river's toxicity in 22 locations (Walker 1990) and it concluded that the Rouge River's ambient water toxicity problems occur only during rainstorms; thus, the river can support aquatic life during dry-weather conditions. This indicates that contaminated sediments do not create an ambient water column toxicity problem for the fishery during dry weather conditions. While this is good news, further work is necessary to determine whether toxicity levels during rainstorms are a result of CSOs, stormwater, or nonpoint source pollution. Any work undertaken to control sources will improve the waterway. One effort funded under Act 307 includes investigation of cross connections to direct discharge storm sewers. The Wayne County Health Department has investigated these direct, unpermitted industrial discharges to the Rouge for three years and will make recommendations on remedial actions in the near future.

Sewer Projects

Several major sewer construction projects are ongoing in the Rouge basin that will enhance the river's water quality. These multi-million dollar projects were being planned and designed when the RAP was adopted. The North Huron Valley/ Rouge Valley Project ($160 million), the Evergreen/Farmington Project ($55 million) and the First Hamilton Connector in Detroit ($37 million) all will provide additional sewer capacity for wastewater from 47 communities in the Rouge basin to the Detroit Wastewater Treatment Plant. These projects will eliminate all dry-weather sewage bypasses throughout the basin and also will reduce the overflow of combined sewage to certain reaches of the Rouge River. It should be noted that the cost figures presented are based on project estimates in 1990.

The Pump Station 2A project and implementation of a flow management plan ($115 million) at the Detroit Wastewater Treatment Plant will provide additional dewatering capacity from the massive interceptor network serving the Rouge basin and will reduce the frequency of CSOs within the basin. It will also eliminate dry weather bypasses within the Detroit collection system.

The Western Township Utility Authority Project ($66 million) will provide additional capacity to communities on the western side of Wayne County. This innovative approach to cross-basin pollution control will direct treated sanitary wastewater from Belleville Lake in the Huron River basin (where total phosphorus loading results in unacceptable nutrient levels) into the Rouge River basin. This treated wastewater effluent will then provide low-flow augmentation in the Lower Rouge basin where inadequate flow is a major problem.

The City of Farmington is actively pursuing funding through the State Revolving Loan Fund Program to separate the remaining combined sewers within its community. This will prevent the raw sewage discharge during rainfall and also will eliminate several separate sanitary sewer bypasses within the collection system.

DNR Regulatory Activity

In recognition of the importance of aggressive regulatory oversight of the individual dischargers on overall water quality improvement, Michigan DNR increased its review of self-monitoring data for direct dischargers within the Rouge basin. The department also conducted intensive wastewater surveys at these facilities to determine the quantity and quality of the existing industrial/municipal direct dischargers and to ensure that the permits for these facilities correctly characterized and limited these discharges.

As a result, enforcement actions have been initiated against industrial dischargers releasing harmful quantities of pollutants into the Rouge River. Enforcement orders also have been issued to all communities in tributaries in the Evergreen/Farmington section, the North Huron Valley/Rouge Valley section and the City of Detroit to assure that needed sewer projects are constructed and properly operated. A sewer construction ban was ordered against Canton Township in 1986 due to raw sewage bypasses from an inadequate separate sanitary sewer system serving portions of the township. Procedures have been implemented in Wayne County and the portion of Oakland County served by the North Huron Valley/Rouge Valley system to ensure that these communities will operate within the contractual capacity allocated to each community within the regional sewage collection system. This will prevent future dry weather bypasses of raw sewage to the Rouge River and will assure that all sewage generated within the basin will receive adequate treatment.

A major requirement of the Rouge River RAP is the control, treatment or elimination of CSOs (SEMCOG 1988). Since 1988, local, state and federal governments have reached agreements and have funded over $500 million in sewer improvements. To force additional controls, Michigan DNR has issued National Pollutant Discharge Elimination System permits to all communities within the basin with CSOs. These permits require the communities to treat the CSOs by the year 2005 in accordance with public health protection dictated by the RAP. This is the first step in implementing of the statewide CSO correction program within the Rouge basin. The communities have contested these permits through the federal court and the Michigan administrative procedures process.

Habitat Restoration

The City of Southfield, in cooperation with the Michigan Wildlife Habitat Foundation, created a sequence of deep pools and shallow riffles by constructing six triangular wing dams in a 0.5 km (0.3 mile) stretch of the Rouge River. These physical modifications were completed in 1988 and have increased the diversity of benthic macroinvertebrates and the quantity of panfish and game fish habitat. Smallmouth bass were subsequently planted and observed in that section of the stream only one year later. The project cost was $8,000 and its success resulted in the City of Southfield receiving the 1988 Clean Waters Award from the Michigan Outdoor Writers Association. Further plans have been made to expand this habitat project by 1.5 km (0.9 mile) in 1991.

It should also be noted that high quality habitat and water is found in the headwaters of the Rouge River. Fisheries biologists from Michigan DNR have found an endangered species—the Redsided Dace—in the headwaters; because this is the only location in Michigan where it is found, special efforts are being taken to preserve and protect this remaining habitat.

Zinc-Contaminated Sediment Remediation

Sediments in the lower Rouge River were found in 1986 to be contaminated with elevated levels of zinc (up to 2,500 mg/kg), which had been discharged in violation of Double Eagle Steel Coating Company's National Pollutant Discharge Elimination System permit. As a result of this zinc contamination and toxicity, a Consent Decree was entered with Double Eagle Steel Coating Company for cleanup costs, damages and penalties (State of Michigan—Ingham County Circuit Court 1986). Rouge Steel Company and USX Corporation, which jointly own Double Eagle, agreed to improve their industrial wastewater treatment process, complete a comprehensive dredging program to remove zinc contaminated sediments, and pay $775,000 in penalties for past violations and $100,000 in damages. Considerable improvements were made in industrial wastewater treatment and have resulted in more than a 99 percent reduction in effluent zinc concentration and loading.

After these improvements were completed, approximately 30,600 m^3 (40,021 cubic yards) of zinc-contaminated sediments were removed from a 1.5 km (0.9 mile) stretch of the river by mechanical dredging. The dredged sediments were placed in the U.S. Army Corps of Engineers' confined disposal facility (Cell No. 5) at Pte. Mouille in southwestern Lake Erie. All dredging and disposal was completed by October 1987 at an approximate cost of $1 million to Rouge Steel Company and USX Corporation.

Where Do We Go from Here?

The Rouge River RAP is designed as a dynamic document. The RAP has identified problems and possible solutions. As results from ongoing studies and/or implemented projects indicate a change in focus or level of effort is needed, this can be incorporated into the document and future RAP recommendations can be changed accordingly.

New Study and Regulation Areas

This dynamic process allows a basinwide approach to investigations and regulations, including such items as the new nonpoint source control program and pending federal stormwater discharge regulations. U.S. EPA recently has made funds available under Section 319 of the Clean Water Act to address pollution from nonpoint sources. Several applications have been received related to the Rouge River, including SEMCOG, which requested funding for: 1) developing an inventory of the open, bulk storage along the shoreline in the shipping channel; 2) developing a geographic information system to identify potential pollution sources and effective utilization of resources; 3) studying the impacts of using wetlands as stormwater retention basins; and 4) studying the impacts of runoff from golf courses, cemeteries and maintained subdivisions on the Rouge River.

Wayne County has received Section 319 funding to complete cross-connection work in one of the most polluted tributaries of the Rouge River. The City of Southfield is also expected to begin similar efforts using Public Act 307 funds. All available funding sources will continue to be examined to pursue actions that further any RAP recommendations. Monitoring programs will continue to evaluate the effectiveness of remedial actions and document improvements, and the RAP will incorporate any results or products of these efforts and reevaluate recommendations accordingly.

The RAP community involvement process was designed to cooperatively resolve environmental issues within the basin. The implications of the new U.S. EPA stormwater regulations, expected to be promulgated in 1990, have been reviewed in draft form and alternative methods to implement the program on a basinwide approach through the RAP process are being evaluated. This should lead to more effective implementation of this program's regulations for the communities.

Implementation: Impediments and Challenges

In general, the RAP is essentially past the planning and study phase and into implementation. Some agencies are more involved in this stage than they were during the planning process. The U.S. Army Corps of Engineers (Corps) has reviewed the RAP goals and developed proposals to assist in accomplishing them. These proposals range from dredging PCB-contaminated sediments in the Middle Rouge and log jam removal to actual construction of CSO retention basins and sewer projects. The assistance of the Corps will help the Rouge efforts; however, most implementation actions must be taken by the communities.

Even as these actions are being implemented, new problems are arising. The planning and development stage of the RAP had a certain amount of self- generated motivation. There were clear, short-term goals to be met. With this effort completed, implementation should follow, but it requires staff, equipment, and funds. A degree

of resistance and a division over the cost of implementation is forming. This has led to the development of "splinter" groups that tend
to address issues with little or no regard for a basin perspective.

Some of this cannot be prevented. However, a review of the RAP
process has shown at least one glaring oversight: no mechanism
was provided to support feedback from the communities to the Executive Steering Committee as implementation proceeds. In many
cases, implementation of the RAP may involve different approaches
for different communities. The existing communication structure
does not readily support sharing success and failure stories, which
is necessary to ensure productive and effective implementation. To
fill this gap and to provide for an internal evaluation, a survey form
is being developed for use throughout the basin.

The survey form will be divided into three parts.

1. The first section will ask Basin Committee members to comment on the ongoing focus of the RAP process. Perhaps their
   present needs and priorities are different from those outlined
   in the RAP. Wherever possible, new needs will be addressed
   and, if necessary, existing priorities will be re-examined.
2. The second section will provide the opportunity for Basin
   Committee members to share their successes and obstacles.
   Successes will receive congratulations and obstacles will receive mutual community support and encouragement from
   across the basin. Hopefully, all the members will gain from
   the others' efforts through this process.
3. The third section will ask the Basin Committee members to
   reevaluate the RAP recommendations in view of their experiences. In this section, committee members can identify
   whether certain recommendations are outdated and need to
   be eliminated or modified, or whether new ones need to be
   considered.

The survey will be issued annually, three to six months before a
Rouge River RAP update is presented to the Executive Steering
Committee. The results of the survey will be used to evaluate progress, applaud outstanding community or interest group efforts,
share information and suggest future directions for the RAP. This

will provide suggested directions for the Executive Steering Committee to act upon.

Conclusion

This dynamic process will continue. The RAP has been a huge cooperative public and private effort to restore the Rouge River. Rather than headlines as mentioned earlier, new headlines should read, "People saw the problems, started working and the Rouge is slowly being restored."

REFERENCES

Cowels, G. 1975. Return of the river. *Michigan Natural Resources.* 44(1):2–6.
Evans, E. 1990. *Rouge River sediment contaminants.* Michigan DNR. Lansing, Michigan. (In Press).
Hartig, J.H. 1984. *Status of water quality and pollution control efforts in the Rouge River Basin.* Michigan DNR. Lansing, Michigan.
Jackson, G. 1975. *A biological investigation of the Rouge River, Wayne and Oakland Counties, May 17 to October 19, 1973.* Michigan DNR. Lansing, Michigan.
Southeast Michigan Council of Governments (SEMCOG). 1988. *Remedial action plan for the Rouge River Basin.* Vol. 1: Executive Summary. Detroit, Michigan.
State of Michigan—Ingham County Circuit Court. 1986. *Consent Decree.* F.J. Kelley, Attorney General vs. Double Eagle Steel Coating Company. Lansing, Michigan.
Walker, B. 1988. Rouge River revival. Michigan Wildlife Habitat Foundation-*Habitat News.* Spring: 1–2.
Walker, B. 1990. *Summary of the Rouge River System ambient chronic toxicity evaluation results from four test periods, January–August, 1989.* Michigan DNR. Lansing, Michigan.
Wayne County. 1990. *Rediscover the Rouge.* Wayne County Parks Division. Westland, Michigan.

Chapter 5

A "Two-Track Strategy" for the Buffalo River Remedial Action Plan

Barry Boyer and John McMahon

> "The remedial action plan, which was developed in cooperation with the Buffalo River Citizens' Committee, is an example of how citizens and government can work together to protect one of our most valuable natural resources."
>
> Thomas Jorling
> Commissioner
> New York State Department
> of Environmental Conservation

Introduction

The Buffalo River historically has been one of the most polluted waterways in the Great Lakes region. In 1968, the U.S. Federal Water Pollution Control Administration reported that "the Buffalo River is a repulsive holding basin for industrial and municipal wastes...It is devoid of oxygen and almost sterile. Oil, phenols, color, ... iron, acid, sewage and exotic inorganic compounds are present in large amounts" (U.S. Department of the Interior 1968). Pollutants moving out of the river affected downstream waters of the Niagara River, Lake Ontario and the St. Lawrence River.

Fortunately, pollutant loadings to the Buffalo River have been greatly reduced in recent years as a result of improved control programs and changes in industrial practices. Historical contaminants still remain in river sediments, however, and combined sewers overflow periodically into the river. Restoring environmental quality in the Great Lakes can best begin with harbors and tributaries like the Buffalo River, where pollutants are most concentrated, and where remedial actions will be most effective if carried out before these pollutants contribute to a lakewide problem.

Development of a remedial action plan (RAP) for the Buffalo River began in 1987, when the lead agency—the New York State

Department of Environmental Conservation (NYSDEC)—appointed a citizen advisory committee to assist in the planning process. Because this was one of the first RAPs initiated in New York state, the nature of the task and the working relationship between the NYSDEC and the Citizens' Committee were initially unclear. Over time, an effective approach evolved that can be described as a two-track strategy for remedial action planning.

In this approach, remedial action planning requires technical analysis and public involvement. The former component, which might be called the "rational planning track," involves the analytical tasks specified in the Great Lakes Water Quality Agreement's general format for remedial action plans:

i.   Identify goals;
ii. Assess impairments by examining information on water quality, sediments and aquatic life in relation to the 14 use impairments defined in Annex 2 of the Great Lakes Water Quality Agreement and the New York State stream classification system;
iii. Identify pollutants or disturbances to determine the cause of impairments;
iv. Identify sources of pollutants or disturbances;
v.   Describe remediation strategy and commitments;
vi. Describe monitoring program measurements that will be used to evaluate whether the remedial actions are proving successful, and whether the indicators of use impairment demonstrate recovery; and
vii. Describe tracking through progress reports and periodic RAP updates.

Rational planning thus involves specifying paths to attain clearly defined, pre-established goals.

In contrast to this rational planning track, strategic planning focuses "on the development of strategic issues which arise as much or more from the values of external stakeholders, perceived opportunities and threats in the external environment, and perceived internal struggles and weaknesses as from a sense of the agency's mission" (Kaplan 1986). From this perspective, the planning process should be fluid, evolutionary, and responsive to the interests of the participants and the larger community. If it is successful, strategic planning will give participants a sense of ownership in the final product because it will incorporate their interests and perspectives.

These distinctions between strategic and rational planning help to describe the development of the Buffalo River RAP, but several aspects of the process require some elaboration. One is the importance of the participants' time horizons. Planning is normally conceived as a discrete event, limited in time to a few weeks or months, and followed by a period of implementation. Because RAP development is a novel, more holistic form of planning that often involves major physical and economic changes in the target area, the time dimension stretches and the distinction between planning and implementation becomes blurred. Remedial action planning is sometimes described as "an iterative process," but it seems more accurate to consider each RAP as "a continuing process" of learning about the ecological interactions within and around a particular geographic area, taking actions to reduce stresses on that ecosystem, and monitoring the system's response. Because natural systems are extremely complex and the knowledge and resources necessary to rehabilitate them are limited, remedial action planning is a process without a clearly defined end point.

The long, indefinite time horizon of RAPs, along with the novelty of the task, suggest that the participants' interests and agendas will be dynamic rather than fixed. Different people enter and exit from the process as it unfolds; relationships develop and change; and shared understandings arise over time. As this occurs, interactions between the RAP agencies and their constituencies may be altered as relationships created by the RAP expand to encompass a broader range of issues. In these respects, interests may be created as well as pursued in the RAP process.

Rational Planning: Defining Impairments, Seeking Solutions

To understand the problems of the Buffalo River and the remedial actions needed to resolve these problems, it is important to review the river's geography, hydrology, and its current and desired beneficial uses as defined by the RAP (NYSDEC 1989).

The Buffalo River Area of Concern falls within the boundaries of the City of Buffalo, New York, near the point where Lake Erie flows into the Niagara River. The Buffalo River extends about 9.6 km (six miles) from the eastern border of the City of Buffalo to its mouth on Lake Erie. The river is used as a transportation channel, passing thorough an industrial area that includes some active sections, many abandoned buildings, several junk yards, and vacant, trash-littered lands. The area gives the appearance of an industrial wasteland.

The primary uses of the Area of Concern are industry, commercial shipping and recreation. Industrial activities include two grain milling firms, two chemical manufacturing companies and an oil company storage terminal. Until the 1980s, a major steel manufacturer, a coke plant, and an oil refinery were located along the river. To provide these firms with adequate processing and cooling water, the Buffalo River Improvement Corporation (BRIC) was formed in the late 1960s. BRIC pumps clean water from Lake Erie to these and other industrial plants along the river. The additional flow provided by BRIC water and the accompanying dilution of pollutants in the Buffalo River produced significant improvements in water quality. However, when industrial plants closed, BRIC activity declined from 454,000 m³/day (120 million gallons per day) in the late 1960s to 68,000 m³/day (18 million gallons per day). Today, industries operate under strict state pollution control regulations, but past operations have created a legacy of contaminated sediments on the river's bottom and abandoned hazardous waste sites along its banks.

The City of Buffalo discharges excess water collected during times of runoff through a combined sewer overflow (CSO) system. The system was designed to collect and transport sanitary sewage and wet-weather flow in the Buffalo area. Although CSOs prevent sewage from backing up into basements and flooding city streets during storms, they discharge untreated sewage into surface waters like the Buffalo River during some storm events, and this is an ongoing problem.

The Buffalo River also serves as part of the region's harbor for Great Lakes shipping. In order to maintain access for commercial freighters, the U.S. Army Corps of Engineers periodically dredges to a depth of 6.7 m (22 feet) below low lake level. Dredging disturbs bottom life, and the bulkheading and dock construction along the river bank have destroyed wetlands and shallow areas that once provided habitat for fish and wildlife.

With the decline of industry and commerce in the Area of Concern, recreational activities such as boating, fishing and swimming are on the rise. Small powerboats travel the river, primarily near its mouth where several marinas and boat mooring facilities have been established. Fishing and swimming are occasionally enjoyed in the area but are restricted by several factors: state advisories against consuming certain species of fish due to toxic substances contamination; a limited number of land access points; a widespread perception that the river is polluted; and the proximity of clean, alternative fishing and bathing sites.

The hydrology of the river is affected by dredging activities and the character of bottom sediments. Dredging slows river flow and increases the volume of backflow from Lake Erie. When flow is high, the river has a "riverine" (unidirectional) character. Under low-flow conditions, the river becomes "estuarine" (with bidirectional flow), and is particularly influenced by lake level variations associated with the passage of storms through Lake Erie and by seasonal thermal differences between lake and river waters. The river and lake waters do not remain separate, but mix at varying rates depending on relative water temperatures.

Studies of bottom sediments show that the river traps all sand particles until its flow exceeds 566 m³/second (20,000 cubic feet per second). This occurs rarely. Finer clay and silt particles are carried downriver during high flows associated with most storms, but fall to the bottom as sediment during normal and lower flows.

The watershed of the Buffalo River has a drainage area of 1,155 km² (446 square miles) and is fed by three tributaries: Cazenovia Creek, Buffalo Creek and Cayuga Creek. The drainage basin provides fish habitats ranging from brook trout habitat in some upper streams to warmwater species habitat in lower urban areas. The creeks receive direct discharge from municipal and industrial treatment plants as well as CSOs. The creeks also receive indirect runoff that contains pollutants picked up from agricultural, suburban and rural land. Despite these inputs, water quality in the tributaries is high. The streams meet New York state's Class A standards (suitable for drinking water), and a 1987 analysis showed that volatile organic compounds were virtually absent from these stream waters.

In general, then, the conditions that are most troublesome result from historic waste management practices such as discharge of persistent toxic pollutants into surface waters, careless disposal of hazardous wastes near watercourses, and construction of combined sewer systems. Reversing the effects of these historic practices will require time and substantial financial and human resources.

Because complete restoration of the Buffalo River will be a lengthy process, the Stage 1 RAP establishes both long- and short-term goals. The long-term goal of the RAP is to eliminate the discharge of pollutants to the Buffalo River. This includes, but goes beyond, the Great Lakes Water Quality Agreement policy of zero discharge of persistent toxic substances. However, the immediate intent of the Buffalo River Stage 1 RAP is to restore and maintain the chemical, physical and biological integrity of the Buffalo River ecosystem in accordance with the Great Lakes Water Quality Agreement. To meet this goal, the RAP seeks to restore water quality so that fish, shellfish and wildlife will survive and reproduce, and recreation can be enjoyed in and on the water consistent with state laws, rules and regulations as they continue to evolve.

Many applicable state legal controls for water quality are incorporated into New York's stream classification system. As indicated in Table 6, most impaired uses that must be addressed in the Buffalo River RAP relate to healthy, normally reproducing populations of fish and other aquatic organisms. Re-establishment of a viable, self-reproducing fishery and its supporting food web and providing a fishery suitable for unlimited human consumption would accomplish most of the goals of the Great Lakes Water Quality Agreement.

Under the New York stream classification system, surface waters like the Buffalo River are ranked according to the desired "best use" of the water resource. The classification takes into account such factors as the character of the bordering lands, stream flow and the past, present and desired future uses of the water. After a public hearing process, the NYSDEC assigns one of the following rankings, based on best use:

Class	Best Use
A	Drinking water
B	Primary contact recreation
C	Fishing and fish propagation
D	Fishing

Each class includes all of the best uses for classes below it, so that an "A" classification would support all desirable uses of the resource. For each designated classification, there is a specific set of standards defining the type and quantity of substances the water can contain and still be used as intended. Both the classifications of particular water bodies and the standards defining acceptable water quality for each classification are subject to periodic public review.

During the development of the Stage 1 RAP, the Buffalo River was classified "D," meaning that it would support fish survival but not necessarily fish propagation, swimming, or a drinking water supply. As part of its regular reclassification review, NYSDEC proposed to upgrade the river to a "C" classification. The citizen advisory committee and other community organizations have taken the position that a "B" or swimmable classification should be given to reflect the rise of recreational uses and the need to accord high priority to combined sewer remediation for the Buffalo River. While this issue has not yet been resolved, the fact that agency officials and public constituencies have engaged in a dialogue concerning the compatibility of the state stream classification system with the goals of the Great Lakes Water Quality Agreement is a hopeful sign that ecosystem objectives will be more explicitly incorporated into state water quality management programs and public consciousness in the future.

Since the Buffalo River Stage 1 RAP was based on a long-term perspective, an implementation strategy needed to be developed that related the various components of remediation to each other, and that sequenced implementation in an orderly fashion. This approach ultimately led to creation of a large flow chart that established functional and temporal relationships among the tasks to be completed to restore the river (see NYSDEC 1989).

The problem areas addressed in the RAP can be separated into monitoring, sediments, point and nonpoint sources, and fish and wildlife habitat. Upgraded monitoring of the river's water and sediments is an essential first step to determine whether applicable criteria are being exceeded and, if so, the source of the contaminants. Monitoring can provide a measure of progress in restoring beneficial uses and a method to identify future targets for remedial action.

TABLE 6. Summary of Impairments, Causes and Sources

Impairment Indicators	Impairment	Likely Cause	Known Sources	Potential Sources
Restrictions on fish and wildlife consumption	Yes	Polychlorinated biphenyls	bottom sediments	inactive hazardous waste sites
		chlordane		bottom sediments
Tainting of fish and wildlife populations	Likely	polynuclear aromatic hydrocarbons	bottom sediments	inactive hazardous waste sites combined sewer overflows
Degradation of fish and wildlife	Likely	low dissolved oxygen[a]		bottom sediments inactive hazardous waste sites combined sewer overflows other point sources other nonpoint sources
Fish tumors and other deformities	Yes	polynuclear aromatic hydrocarbons	bottom sediments	inactive hazardous waste sites combined sewer overflows
Bird and animal deformities or reproduction	Likely	polychlorinated biphenyls	bottom sediments	inactive hazardous waste sites
Degradation of benthos	Yes	none identified	bottom sediments	
Restriction of dredging activities	Yes	metals and cyanide	bottom sediments	inactive hazardous waste sites combined sewer overflows other point sources other nonpoint sources

Eutrophication or undesirable algae	No	
Restrictions on drinking water consumptions or taste and odor problems	No	
Beach closings	No	
Degradation of aesthetics	No	
Added costs to agriculture or industry	No	
Degradation of phytoplankton and zooplankton populations	No	
Loss of fish and wildlife habitat	Yes	physical disturbance	bulkheading dredging steep bank slopes

Note: Impairments, causes, and sources were identified for the Buffalo River by the RAP team.
[a] River channelization is also a potential factor.

Control over contaminated sediments is the least developed program area, not only in the Buffalo River Area of Concern but
throughout the Great Lakes. Initially, two parallel tracks need to
be followed. First, sediment criteria must be developed so that contamination "hot spots" can be identified and numerical goals for
remediation are established. Second, the dynamics of sediment
transport in the river must be more fully documented to support
an informed decision about the adequacy of natural "armoring" or
sediment burial as an acceptable alternative for some reaches of the
river. Once the basic choice is made between burial and removal of
contaminated sediments, more detailed evaluation of particular
treatment or destruction options can be completed.

Recent initiatives in the Great Lakes basin will facilitate these
local efforts to cleanup contaminated sediments. The Buffalo River
has been included as a demonstration site under a federal program
for Assessment and Remediation of Contaminated Sediments
(ARCS), and a "mini-mass balance" study is underway to map the
location and movements of the river's contaminated sediments.
Other regional efforts, such as the IJC's publication of guidance
documents on sediment assessments and the U.S. Environmental
Protection Agency's program to develop sediment criteria, will fill
in further pieces of the sediment puzzle. Experience in remediating
underwater sites through federal and state Superfund programs may
also provide valuable information.

Before a major sediment cleanup program can be launched, however, it is essential to eliminate sources that could recontaminate
the river bottom. Two significant sources of ongoing inputs are
known: approximately three dozen inactive hazardous waste sites
within the Buffalo River watershed, and several permitted point
source discharges to the river and its tributaries. The inactive waste
sites are being addressed under New York's "superfund" program,
on a schedule which requires cleanup to be completed by the turn
of the century.

Since the primary industrial discharges to the lower river include
noncontact cooling water, these sources require only routine compliance monitoring and possible revision of permits as standards
change. CSOs, however, are a more intractable problem that will
require study and action to resolve. The first step in addressing the
CSOs is to prepare a computer model of the main sewer lines affect-

ing the river. With that model, possible remedial alternatives and their feasibility can be identified.

In addition to these known pollutant sources, it is possible that other point or nonpoint sources will be identified as monitoring and remedial actions progress. The remedial strategy specifically requires assessment of this possibility and control over all input sources before sediment remediation reaches the implementation stage.

Finally, fish and wildlife habitat restoration will move forward on a parallel track. Ideally, assessment of habitat needs and specification of improvement plans could wait until remedial actions were well along, so that fish and wildlife specialists can take full account of the river's altered physical and chemical characteristics. However, economic and social conditions in the Area of Concern dictate a different approach. Waterfront revitalization planning is underway at the city and regional levels and the pace of private development along the river also seems to be increasing. Delaying habitat assessment for several years during this period of changing economics and land use regulation might mean that it would be impossible to acquire large land parcels for habitat improvements (e.g. restoration or creation of wetlands) because the desirable areas had already been taken over by inconsistent developments or had become prohibitively expensive. Thus, the RAP remedial strategy calls for early assessment and acquisition of habitat improvement areas.

Once this general schematic of remedial actions was in place, the first steps needed in each program area could be specified, and commitments could be made by the responsible agencies. These initial remedial actions, as specified in Table 7, include ongoing program activities and new resource commitments. Most initial program targets were met during the first year of RAP implementation, following the submission of the Stage 1 RAP to the IJC in November 1989.

A 12-member Remedial Advisory Committee has been appointed by the NYSDEC to review the agencies' progress in meeting the targets established in the Stage 1 RAP, to update that document, and to lay the groundwork for eventual preparation of a Stage 2 RAP outlining specific steps needed to restore the Buffalo River. The Remedial Advisory Committee meets on a quarterly basis and has

TABLE 7. Summary of Remedial Options Recommended by the RAP Team

Remedial target	State commitments to date	Objective	Recommendation
Bottom sediments	Correct impairment to fishery and aquatic life caused by contaminated sediments.	—develop model of sediment, flow and deposition to determine armoring removal potentials —develop sediment criteria —assess river sediments based on criteria to determine where remedial work is needed	—develop requirements for a model to predict scouring and deposition —urge EPA to develop national sediment criteria
Fish and wildlife	Improve habitat in and along the river.	—assess conditions and potential for improvements —develop habitat improvement plan —acquire necessary land and carry out improvements	—develop plan for assessment of habitat conditions and improvements by March 1990
Water quality monitoring	Ensure all sources have been addressed. Determine impacts of low dissolved oxygen on the fishery.	—establish automated sampling station —develop models to relate contaminant levels to their potential for harming fish or aquatic life. —carry out intensive dissolved oxygen study	—establish flow-activated sampling station by 1990 —measure dissolved oxygen in river to determine if it is lacking and then determine causes of decreased levels

Inactive hazardous waste sites	Prevent sites from contributing contaminants to the river.	—continue ongoing program for remedial work —conduct nine investigation studies at sites in 1990	—investigate each of nine sites in 1990
Other nonpoint sources	Prevent adverse effects on the river.	—use stream monitoring to determine if nonpoint sources are significant sources —determine which nonpoint sources require remedial actions	
Municipal and industrial wastewater facilities	Ensure no significant contribution to impairment of fishery or aquatic life.	—renew permits incorporating current technology- and water quality-based limits —monitor discharges and enforce compliance if necessary	—continue permitting program with renewals every five years
Combined sewer overflow systems	Ensure no significant contribution to impairment of fishery or aquatic life.	—develop modeling system for improvements to increase flow	—BSA[a] is developing and evaluating model of CSO system (required under SPDES permit) to be completed in 1990
Other point sources	Ensure no significant contribution to the impairment of fishery or aquatic life.	—identify point sources of concern	

Note: These recommendations form part of the Buffalo River RAP (NYSDEC 1989).
[a] Buffalo Sewer Authority

set up a committee structure to enable participants to monitor all relevant program areas.

Strategic Planning: The Social Context of RAP Development

With the benefit of hindsight, it is easy to overemphasize the rational, orderly character of the process leading to the creation of a Stage 1 RAP for the Buffalo River. Especially during the early stages of the process, the nature of the task was obscure, the working relationship between the lead agency and the citizens' group was changing, and the need to generate political and community support was not generally understood. At the outset, there was little guidance from either the IJC or the U.S. Environmental Protection Agency regarding either the content of the document or the role that citizen groups should play in developing the plan.

Within the IJC, there seems to be broad support for the idea that public participation is essential to the ultimate success of RAPs. Descriptions of the process by involved IJC officials speak broadly of the need to develop a teamwork approach toward writing a RAP (Hartig and Vallentyne, 1989) and to generate political support for implementation of the remedial measures contained in the plan (Hartig and Thomas, 1988). However, these favorable descriptions of public participation give scant detail as to how consensus will be gained once everyone has been brought around the table, or why it should be assumed that any consensus resulting from the process will necessarily complement the ecosystem goals of the Great Lakes Water Quality Agreement.

At the same time, the IJC's enthusiasm for public involvement does not seem to be shared equally by all parties who bear responsibility for preparing and implementing the plans. Guidelines developed by a contractor for the U.S. Environmental Protection Agency's Great Lakes National Programs Office, for example, assert that "although RAPs will be available to the public, they will not be written in a style which seeks to educate and inform the layman as is an environmental impact statement" (Science Applications International Corp. 1985). Fortunately, the state and provincial governments seem to have ignored this advice in their RAP programs.

The NYSDEC began the Buffalo River RAP with a strong commitment to public involvement. The Buffalo River Citizens' Committee was appointed by the NYSDEC in response to citizen initiative, and it includes a broad spectrum of community interests and organizations as requested by local citizen representatives. A special staff person was hired in the NYSDEC's regional office as Citizen Participation Specialist to facilitate public involvement, and further support for public participation was provided by the agency's headquarters staff. Despite these formal commitments, however, the relationships between the agency and public still had to be worked out in practice by the RAP participants. In Buffalo, as in other remedial action planning efforts, "the first task is to find a way to think about political participation" (Milbrath 1965).

For the Buffalo River RAP, it seems useful to analyze the public participants' roles within the framework of "regulatory culture" developed by Meidinger (1987). According to Meidinger, specific fields of regulation like RAPs generate ongoing relationships among key actors and institutions that can be viewed as "regulatory communities." Within these communities, actors pursue both material interests and ideal visions. The net result is a set of shared understandings that comprises a distinctive culture:

> As in ordinary communities, members of regulatory communities have ongoing relationships with each other. In these relationships, they both pursue their own, often inconsistent interests and struggle to define a shared vision of the collective good. Because they live significant parts of their lives with each other, members of the community frequently influence each other, act with reference to each other, and desire each other's respect. Therefore, as well as being arenas for the pursuit of preexistent interests, regulatory communities appear to have the capacity to be "constitutive"—that is, to be forums in which appropriate individual and collective behavior (and interests) are defined and redefined (Meidinger 1987).

This formulation fits the Buffalo River experience in two respects. First, it highlights the significance of relationships in determining the nature and success of the endeavor. Because remedial action planning requires a more holistic ecosystem approach than the existing, bureaucratically specialized fields of environmental regulation, RAP development cannot be dominated easily by a particular person, bureau or discipline. Moreover, the lack of detailed official guidance during the crucial early stages of RAP development gave additional importance to the participants' shared understandings. With only sketchy blueprints to follow, the skills and visions of the construction crews become highly significant.

A second advantage of the regulatory culture framework is that it emphasizes the interplay between material and nonmaterial incentives. In the Buffalo River RAP, while personal and economic agendas were significant, the process was not solely driven by them. Rather, the ideals and ideologies of the participants played a major role in shaping the document and creating other nonprofit volunteer organizations to carry forward the work of the RAP.

The concept of regulatory culture does not, however, provide a detailed framework for analyzing how conflicts of interest or of vision are resolved. For this purpose, a focus on strategies is appropriate.

According to Bardach (1977), statutory implementation is "a process of strategic interaction among numerous special interests all pursuing their own goals, which might or might not be compatible with the goals of the policy mandate." From the perspective of the program proponents and particularly for the lead agency, implementation becomes a matter of assembling the necessary program elements internally, and using persuasion and bargaining to induce others—including other offices within the same agency—to contribute necessary program elements (Bardach 1977). Those more peripherally involved, on the other hand, may maneuver to minimize their own contribution, to claim credit or avoid blame, or to divert the new program toward their own goals. These interactions give rise to a series of patterned strategies or "games" that determine how a new program will be implemented or whether it will be implemented at all.

Bardach's framework captures an important fact of the RAP process for the Buffalo River: virtually no new program resources were made available from either the federal or state levels to develop the plan. Thus, the participants were forced to cannibalize most of the necessary resources such as staff time, expertise, data and support for implementation from other ongoing activities.

Time was an important factor in determining how the implementation strategies would be developed. As previously noted, many participants in the RAP process adopted a long-term perspective based on the understanding that it would take many months to prepare an acceptable Stage 1 RAP, and that this document would be only the first step in a much longer process to actually restore the river's degraded ecosystem. Other participants who were unwilling or unable to make this extended commitment dropped out as the process progressed.

Lengthening the time horizon of the process may have several advantages. For example, it can avoid or postpone showdowns on contentious issues, as illustrated by several situations that arose during development of the Buffalo River RAP. Time can also allow enough interaction among key participants to permit convergence of ideas. This function became apparent during one of the rare lapses in communication between the Citizens' Committee and the agency RAP team. During summer 1988, the chief member of the writing team became unavailable for more than a month as a result of vacation schedules and urgent crises. The Citizens' Committee and key officials from the NYSDEC did not meet during this period. The Steering Committee was essential to the planning process because it provided the forum in which disagreements over the embryonic RAP were discussed and consensus frequently emerged. The Citizens Committee met intensively during this period because, under the schedule urged by agency officials, the RAP would move into draft form very quickly during late summer. Thus, it appeared to the public participants that this was their last chance to propose specific remedial actions for the final document. In a series of marathon sessions, the Citizens' Committee hammered out a lengthy list of detailed remedial actions that it wanted incorporated into the plan.

When the Steering Committee eventually reconvened, the NYSDEC had considerable difficulty with many of the Citizens' Committee's recommendations. Some were at a level of detail that the department felt was more appropriate for a workplan than for a RAP; others depended on data that had not yet been collected or decisions that had not yet been made; a few stretched, if not exceeded, the bounds of the NYSDEC's authority. Because the recommendations had been endorsed by the full Citizens' Committee, however, the public representatives did not feel that they could simply scrap most of the detailed recommendations that everyone had worked so hard to create.

A series of lengthy Steering Committee meetings ensued and two measures eventually were worked out to resolve the impasse. The "Remedial Strategy Schematic" flow chart previously described was devised to show the interrelations among the various components of the RAP and the time sequence in which major future actions would be taken. In addition, the detailed Citizens' Committee recommendations that the NYSDEC had been unable to accept were carried forward into the final document, carefully hedged with brackets and labeled as developed by the Buffalo River Citizens' Committee rather than NYSDEC proposals. Thus, one ironic result of proceeding on the "fast track" during the summer was more delay as well as less consensus than might have emerged under normal procedures.

A second illustration of the advantages of a long time horizon in minimizing conflict concerned the question of legislative authority. From the outset, the leadership of the Citizens' Committee wanted the RAP to highlight the shortfalls in budgetary allocations and statutory authority that would prevent prompt restoration of beneficial uses in the river. The NYSDEC writing team was equally reluctant to go on record that they needed more money and power to remediate the river. This resistance may have come from a belief that a sufficient factual case had not been made for legislative action, or from a desire not to get embroiled unnecessarily in the various internal and external clearances and conflicts that a request for legislation would entail; perhaps other factors were involved as well.

In any event, the Citizens' Committee proposed a series of drafts on legislation that were carefully critiqued by NYSDEC personnel in Steering Committee meetings. Ultimately, a resolution was postponed. The RAP's final draft contains a section written by the Citizens' Committee that notes a number of areas in which future legislative intervention may be necessary, but also acknowledges that those problems need not be resolved immediately. In the meantime, implementation can move forward under existing authorities. As this example indicates, working relationships are more likely to remain stable and productive when at least some conflicts can be put off to another day.

The shift in conception from a one-shot work plan detailing specific remedial actions to a longer-term strategy of implementation in which decision points are spread over time seems likely to maximize incentives for cooperation. Axelrod's (1984) research in game theory demonstrates that cooperation is most likely to evolve under conditions that approximate an "iterated prisoner's dilemma," or imagery that is intuitively appealing to anyone who has spent interminable hours in windowless meeting rooms haggling over the technical minutiae of a RAP. The classical prisoner's dilemma resembles a remedial action planning process in that the biggest payoff for all players (and perhaps for the environment as well) is likely to come from mutual cooperation. At the same time, various players or constituencies may be tempted to defect from the cooperative effort if it appears that they can reap some benefits from defection without fear of retaliation (Axelrod 1984).

According to Axelrod (1984), one of the keys to cooperative behavior under these circumstances is the iterated nature of the game, which means that it continues through repeated rounds so that one party's defection in one round can be punished by the other's retaliation in round $n + 1$—the common and winning strategy generally known as "tit for tat." Thus, one of the simplest, most powerful actions to promote cooperation in this social setting is to "enlarge the shadow of the future" by making clear that there will be future rounds of the game in which the players will retain the capacity to reward and punish each other (Axelrod 1984). By contrast, "itinerant players that are here today and likely to be gone tomorrow often will not be treated as potentially cooperative players."

In many environmental controversies, players can be considered itinerants. The classic local dispute over threats to a neighborhood from a leaking toxic waste dump, for example, may have temporary players on both sides. The citizens' or homeowners' group for the most part wants only to be relieved of the health threats and financial losses associated with the particular dumpsite. Once the residents have been removed from the problem area, or the problem has been removed from their neighborhood, they are not likely to have any future involvement with the responsible agencies.

On the agency side, the major players also may well be short-timers. Most government offices have a fairly rapid turnover of personnel, especially when the job in question involves frequent contact with scared and outraged citizens (Levine 1982). Thus, it is perhaps not surprising that commentators who view toxic waste controversies from a "victim" perspective are those who frequently advocate confrontation and political pressure as the only viable strategies to seek redress (Montague 1989). At least some RAPs seem to be dividing between "itinerants" and "local citizens," with each group portraying differing perceptions as to the payoffs for cooperative behavior.

Within the Buffalo River Citizens' Committee, adoption of a longer time horizon made it possible for key citizen participants to develop strategies leading toward accomplishment of long- and short-term goals. These strategies can be divided into access, accountability, and constituency building. All strategies were pursued simultaneously, in a spirit of improvisation and experimentation, by different individuals and subcommittees of the citizens' committee. In addition to the citizen participants, many active agency personnel also were supportive of public involvement strategies, though perhaps more constrained in their actions as a result of their official roles and responsibilities.

Access

From the outset, it was apparent that the Buffalo River RAP needed to focus on questions of access, in two senses. First, as in any process of public involvement in decisionmaking, the affected publics needed to have access to relevant information and key decisionmakers. Second, and probably more important, the public needed to have access to the river and a reason to want to go there. Throughout this century, most of the access points to the lower Buffalo River had been given to private industry. Above the mouth of the river where a marina, a condominium development, and a naval park offered some opportunities for walking along the river, the only public access were two lift bridges and a few streets that ended at the river. Even worse than the lack of physical access was the mindset in much of the surrounding community, which had long ago defined the Buffalo River as an industrial sewer, beyond consideration as a recreational resource. When the RAP process began, the Buffalo was "a forgotten river," as one of the leaders of the Citizens' Committee described it.

In fact, the Buffalo River has considerable potential for active and passive recreation. Many polluting industries closed during the recession of the early 1980s and this fact, coupled with improved pollution controls at the remaining plants, brought greatly improved water quality. Fish and bird populations were re-establishing themselves in the river, and natural revegetation of the shoreline created surprisingly pastoral settings in the midst of a large city. Large abandoned tracts of formerly industrial land opened up possibilities for new private and public uses.

Citizens' Committee members with backgrounds in community organizing soon understood that one of the most urgent tasks was to raise public consciousness about the river's possibilities, followed by establishment of a literal "beachhead" on the river. In cooperation with NYSDEC, a series of activities were initiated to increase the river's visibility. Slide shows depicting attractive scenes along the river and describing the work being done to restore it were created and widely shown to community groups. A series of public meetings inviting residents of the surrounding neighborhoods to share their visions and frustrations for the river were held at the beginning of the RAP process. Walking tours of sites along the river, focusing on the need to clean up pollution sources, were held at several points during RAP development, and a small boat regatta was organized. A mailing list was created from these early activities and a newsletter was sent periodically to interested citizens. Press and broadcast coverage of RAP activities was also helpful in increasing awareness and interest.

Aided by the example of the Rouge River Rescue program in Michigan, which was discussed at a conference sponsored by Great Lakes United, the RAP team began to seek sites along the river that could be acquired for public access or "beachheads." In cooperation with NYSDEC, the Citizens' Committee inventoried land along the river and identified a large tract with an abandoned warehouse as a prime target for acquisition. Plans were developed to create a park and an environmental education center on the site, and funds were appropriated in the city budget. Annual spring "cleanup days" have been held at the site, when citizen volunteers removed trash and debris. While the acquisition of this property has recently encountered legal and political obstacles, the effort invested has been worthwhile as a means to involve the public in a visible, tangible commitment to restoration of the Buffalo River.

In addition to development of the environmental education center and park, two other efforts to provide physical access to the river have helped to generate public support for the RAP's goals. Railroad land along the river was proposed for abandonment, and through a cooperative effort of the Citizens' Committee and NYSDEC, the state initiated the process to acquire several hectares for future development as a boat ramp. A city councilman whose district includes much of the lower river also has successfully advocated a local ordinance requiring a 7.6 m (25 foot) setback for new development projects so that visual access could be preserved and opportunities for future public access enhanced.

These physical access projects have contributed to the goals of the RAP in several ways. By translating the abstractions of "restored beneficial uses" into concrete physical terms that are attractive to the community and by providing short-term, potentially achievable goals, they have given citizens and local political leaders something to rally behind while the technical work of pollution abatement moves forward. In the long run, enhancements in public access promise to create new user groups who will become advocates for the river and who will protect it from future degradation. If the environmental education center becomes a reality, the public's "beachhead" on the river may also promote understanding among future generations about the need to harmonize economic development with the preservation of natural ecosystems.

Parallel to these efforts for providing physical access was a continued commitment that public participants in the RAP process had adequate access to information and decisionmakers. Several factors contributed to the Citizens' Committee's capacity to get and use relevant information. One of the most important factors is the ecosystem approach built into the RAPs. While this approach creates technical frustrations because it is not well defined and much of the information needed to determine the causes of impaired uses may not exist, it does offer one significant advantage: broad, interdisciplinary focus of the ecosystem approach guarantees that no single bureau or office will have a monopoly on relevant data. As a result, developing an adequate understanding of even a small Area of Concern like the Buffalo River requires a collaborative effort in which many individuals who have pieces of information try to fit them together into a coherent picture.

There is some potential for gridlock if a key actor plays an obstructionist game, such as stonewalling the process or making only a token effort (Bardach 1977). But it also can create a productive atmosphere to share among equals and to solve problems. For the most part, the latter dynamic prevailed during the development of the Buffalo River RAP. Several characteristics of the community and the Citizens' Committee helped to create that atmosphere. Several agency offices—whose staff contributed substantial amounts of time, energy and information to the RAP—have offices in Buffalo, including NYSDEC's regional office, the U.S. Army Corps of Engineers' Buffalo District, the City of Buffalo Division of Planning, the Erie and Niagara Counties Regional Planning Board, and the Erie County Department of Environment and Planning. Several local research institutions, such as the State University and State University College at Buffalo and the Roswell Park Cancer Research Institute, also provided technical personnel who contributed to the project. The fact that all viewpoints were represented within the local community made it easy for participants to interact. The information sharing probably would have been much different if the actors had to be brought together from long distances away.

In addition to these community resources, Buffalo was fortunate to have a cadre of volunteers who had little formal background on technical matters but were nonetheless willing to contribute their services in support of those who were performing technical work for the RAP. Through these volunteers, the Citizens' Committee acquired data describing inactive hazardous waste sites in the watershed, and created a computerized data base listing all waste sites containing chemicals that impair uses. Thus, the public participants became part of the effort to compile and analyze data, rather than passive consumers of information served up by the lead agency.

Public access to information will be maintained during RAP implementation through several channels. The U.S. Environmental Protection Agency is strengthening the technical infrastructure by including local researchers in the sediment studies under the ARCS program. In addition, the Remedial Advisory Committee previously described will involve the community in reviewing implementation and updates to the RAP. Through these channels, as well as through the formal and informal networks that have developed around the

RAP, it should be possible to maintain a continuing flow of information. As information is shared and evaluated, the embryonic "regulatory culture" that has grown up around the early stages of the RAP may be elaborated and strengthened.

Accountability

Access to decisions has little utility if the process does not assure accountability in the implementation of those decisions. This accountability must include assignments of responsibility and short-term commitments that measure progress toward long-term goals.

The need for clear assignments of responsibility became evident early in the development of the Buffalo River RAP, when confusion grew about the role of the Citizens' Committee in preparing the draft plan. Eventually, a "Steering Committee" structure was arranged in which drafting was completed primarily by the regional NYSDEC office with input from NYSDEC headquarters and the Citizens' Committee. The overall plan, the timing of various stages and initial approval of drafts were worked out initially in the steering committee (comprised of NYSDEC headquarters and regional personnel; two chairpersons of the full Citizens' Committee, and the chairs of the Public Outreach, Technical and Long-Range Goals committees). This system generally worked well, although it was frequently difficult for the Citizens' Committee representatives to find enough time to participate fully in Steering Committee meetings in addition to their regular jobs and other responsibilities to the RAP process. Designated alternates might have improved the continuity of steering committee deliberations and distributed the workload more efficiently.

In the initial round of RAP implementation, accountability is being fostered by incorporating a series of precise short-term commitments, coupled with an annual review process keyed to the state budget year and involving public participation through the Remedial Advisory Committee. Since this system only recently was established, it is not clear how successful it will be. Much remains to be done to achieve the goals of the RAP. Commitments made thus far have been relatively modest in dollar terms, and generally have fit easily within the mandates of NYSDEC's existing programs; mostly, they have involved "further study" rather than concrete remedial action. This probably will not persist for long: hard decisions must be made and implemented, some of them involving agencies like the U.S. Environmental Protection Agency and the Buffalo Sewer Authority, which have been bystanders to the process thus far. As Bardach (1977) notes, "The policy adoption process wrests from participants verbal support for a proposal, whereas the policy implementation process demands the contribution of real resources. To put it another way, talk is cheap but actions are dear."

Constituency Building

As the difficulty and expense of RAP implementation increase, it will become more important to foster the constituency support necessary to put in place essential program elements. Constituency support for the RAP's goals will have to become broader and deeper, and work must proceed along many fronts: neighborhood residents, public interest groups and other interested members of the general public; specialized publics, such as industrial and recreational user groups; responsible agencies and their relevant departments, bureaus, and offices; developers, planners, and development authorities; and political leaders at all levels of government.

In the early stages of RAP implementation, constituency building has coalesced primarily around two new organizations spawned by the RAP. One, the Buffalo River Environmental Research Association, includes academics and other technically-oriented people who are engaged in research activities that may involve work on or about the Buffalo River. The group, loosely organized as an unincorporated association, includes some individuals who were involved in RAP preparation and some graduate students and faculty members who had not previously participated. In addition to sharing support, ideas and information about research related to the river, the association contains a number of individuals who are willing to provide technical input into public policy decisions affecting the Buffalo River, such as the recent public hearing on reclassifying the river under New York's water quality statute. Members have also participated in some nontechnical activities that will help to build constituency support, such as commissioning a poster featuring attractive photos and drawings of the Buffalo River. The association meets monthly during the academic year.

Another constituency building organization, with some member-ship overlap with the Buffalo River Environmental Research Asso-ciation is the Friends of the Buffalo River. Like the association, the friends organization grew from a core group of participants in the Buffalo River Citizens' Committee. It took a different organiza-tional form, however, as a tax-exempt charitable corporation de-voted to the environmentally-sound revitalization of the Buffalo River and its surrounding neighborhoods. The organization thus is heavily involved in community development issues such as the creation of a park and environmental education center along the river, the rerouting of road networks in south Buffalo, preservation of industrial heritage sites along the river, and planning and zoning activities in the region. Its board of directors reflects this broader focus and includes representatives from academia, community or-ganizing, business, planning agencies and local government. The Friends organization has approximately 50 paid members, offices donated by a community organization, and governmental support in the form of a consulting contract. A staff person was hired under the contract to work on plans for the environmental education cen-ter. Friends of the Buffalo River have held two annual meetings, both attended by the city's mayor.

Whether this organizational momentum can be maintained and increased is unclear. It will be essential to maintain continuity in leadership, and to recruit new members and future leaders. Re-sources for maintaining the organizations over time remain uncer-tain, and the success of some of the group's major projects is far from assured. Still, the prospects are encouraging for building the goals of the RAP into the life of the community, and in effect modi-fying its culture to re-define the Buffalo River as a valued resource rather than an embarrassment and a health hazard.

As RAPs like Buffalo's move from drafting to implementation, it is timely to ask whether the IJC and other interested organizations should be doing more to encourage and facilitate public involve-ment in the restoration of Areas of Concern. Workshops, confer-ences and opportunities for public input such as the hearings at the IJC's 1989 Biennial Meeting in Hamilton are useful, but provide limited tools to transfer knowledge and evaluate progress. Two pos-sibilities may deserve further investigation.

The IJC's process of reviewing RAPs for completeness and

sufficiency is not balanced toward technical and scientific issues. Some reviewers are concerned about whether there has been adequate public participation in the process, but to date there has not been any detailed guidance or feedback to the jurisdictions with regard to what constitutes a good public involvement program. The IJC's technical criteria for RAPs, such as the Listing/Delisting Guidelines for Areas of Concern, also need further work and elaboration, however, they are far superior to any comparable guidance that has taken place with regard to the public's role in the process.

A second and related point is that the IJC as currently organized has little capacity to provide either detailed general guidance or specific feedback on public involvement in the RAPs. Moreover, little academic research on these topics seems to be taking place. Yet, public interest and public involvement in the restoration of the Great Lakes is growing, as evidenced by the outpouring of public concern voiced at the IJC's 1989 Biennial Meeting (IJC 1990). Perhaps it is time for the IJC to consider modifying its organization to reflect this change. It may be time to add a new Public Advisory Board to the Water Quality and Science Advisory Boards that is charged with the responsibility of fostering relevant research and demonstration projects, and conducting and evaluating programs of public involvement dedicated to protection of the Great Lakes.

REFERENCES

Axelrod, R. 1984. *The Evolution of Cooperation.* New York: Basic Books, Inc.

Bardach, E. 1977. *The Implementation Game: What Happens After a Bill Becomes Law.* MIT Press, Cambridge, MA.

Hartig, J.H. and R.L. Thomas. 1988. Development of Plans to Restore Degraded Areas in the Great Lakes. *Environmental Management* 12:327–347.

Hartig, J.H. and J.R. Vallentyne. 1989. Use of an Ecosystem Approach to Restore Degraded Areas in the Great Lakes. *AMBIO* 18:423–428.

International Joint Commission (IJC). 1990. *Fifth Biennial Report on Great Lakes Water Quality. Part I.* Windsor, Ontario, Canada.

Kaplan, T.J. 1986. The Narrative Structure of Policy Analysis. *J. Policy Analysis and Management* 5:761–778.

Meidinger, E. 1987. Regulatory Culture: A Theoretical Outline. *Law and Policy* 9:355–386.

Milbrath, L.W. 1965. *Political Participation.* Rand McNally & Co., Chicago, IL.

New York State Department of Environmental Conservation. 1989. *Buffalo River Remedial Action Plan*, Albany, NY.

U.S. Department of the Interior Federal Water Pollution Control Administration, August 1968, *Lake Erie Report*, Washington, D.C.

Chapter 6

Dredging Up the Past: The Challenge of the Ashtabula River Remedial Action Plan

Julie A. Letterhos

> "I strongly believe that local involvement is the key to successful development and implementation of a RAP. The Ashtabula River RAP process has provided a forum for the local community, state and federal agencies, and legislators to resolve conflicts, to promote cooperation and concentrate efforts to achieve the common goal of removing contaminated sediments from the Ashtabula River."
>
> Richard Shank
> Director
> Ohio Environmental Protection Agency

Introduction

The title of this chapter rather facetiously identifies "dredging up the past" as the major challenge facing cleanup programs for Ohio's Ashtabula River Area of Concern. In reality, this is literally the obstacle that must be addressed. Sediments in the Ashtabula River have been contaminated with PCBs, metals and a myriad of organic chemicals from industrial dischargers, spills and leachate from hazardous waste disposal sites. Fields Brook, a tributary to the Ashtabula River and part of the Area of Concern, was declared a Superfund site under the U.S. Comprehensive Environmental Response, Compensation, and Liability Act in 1983. As with many formerly thriving Great Lakes port cities, the area's faltering economic base of heavy industry and commercial shipping is being bolstered by development of a recreation industry. Unfortunately, accumulated contaminated sediments are restricting development of the recreation industry and creating a health hazard. Although the discharge of some pollutants is still occurring, most direct inputs are largely controlled and not of sufficient volume to cause recontamination of the Area of Concern if the polluted sediments are removed. In order to achieve full restoration of the beneficial uses in the Ashtabula River, the only recourse is to "dredge up the past." This is the main focus of the Ashtabula River remedial action plan (RAP).

Background and Definition of the Problem

The Ashtabula River is geographically the smallest of Ohio's four Areas of Concern, but it has the greatest variety and concentrations of toxic substances, particularly organic chemicals. Located in extreme northeast Ohio, it includes the lower 3.2 km (two miles) of the Ashtabula River, the outer harbor, adjacent Lake Erie nearshore and Fields Brook. A small tributary named Strong Brook is a potential source of pollutants (Figure 11).

In the 1940s and 1950s, an industrial complex was developed on Fields Brook and along the Lake Erie shoreline east of the river mouth. Many industries used byproducts of neighboring companies to support their own operations and to keep costs down, creating a concentrated cluster of chemical companies. Adding to this cluster were power generating plants, coal docks, rail yards and wastewater treatment plants. Runoff from a scrap yard that salvaged transformers and a long-undetected leak from one transformer at a chemical company were the major sources of PCBs. PCB concentrations as high as 518 mg/kg have been found in Fields Brook sediment (CH$_2$M Hill 1985). A number of closed or abandoned hazardous waste landfills have contaminated groundwater and are potential sources of pollutants to the surface waters of the Area of Concern.

Many chemical companies on Fields Brook handled chlorine and chlorinated organic compounds. Segments of Fields Brook were devoid of biological communities because of the toxic effect of this chlorine. Local residents contend that laundry was bleached as it hung on the line and a visible mist often covered the area. This situation no longer exists, but some areas still smell strongly of solvents when disturbed. Fields Brook was placed on the National Priority List for cleanup under Superfund in 1983 because of the possibility for direct contact with the sediment by humans, the movement of contaminated sediment to the Ashtabula River and Lake Erie, the potential impact on the public water supply, and the release of hazardous materials from the sediment.

The U.S. Environmental Protection Agency (EPA) and the U.S.

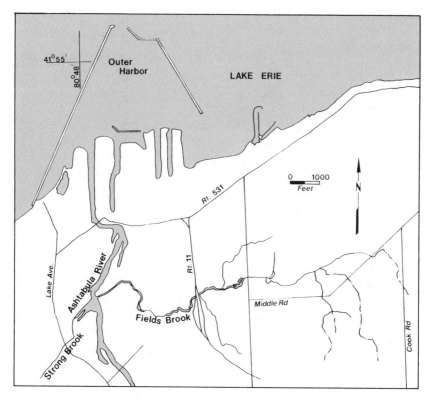

Fig. 11. The Ashtabula River Area of Concern

Army Corps of Engineers have conducted many tests on the pollution characteristics of the Ashtabula River and Fields Brook since 1971. Sediments in the river channel have been classified as heavily polluted, nonpolluted, moderately polluted, hazardous, and toxic. Tests performed in 1982 resulted in the removal of the hazardous designation, but sediments in the turning basin were still toxic due to PCB concentrations greater than 50 mg/kg (Aquatech 1983). Based on the most recent data from bulk chemical analyses, elutriate tests, and bioassays, the sediments of the turning basin and an area immediately downstream of Fields Brook are classified as toxic based on U.S. EPA guidelines; most of the river channel sediments are classified as heavily polluted and the outer harbor sediments are nonpolluted to moderately polluted. The toxic sediments must be disposed of according to the regulations of the Toxic Substances

Control Act (TSCA), which requires disposal in a specially constructed landfill, incineration or a third alternative subject to the approval of U.S. EPA. The heavily polluted material must be contained at a conventional confined disposal facility and all nonpolluted to moderately polluted sediments are suitable for open lake disposal or unrestricted upland use.

According to U.S. EPA guidelines (U.S. EPA 1977), Ashtabula River sediments are heavily polluted with mercury, lead, nickel, zinc, chromium, cyanide, arsenic, barium, copper and cadmium. Organic compounds detected at elevated levels include PCBs, hexachlorobenzene, other chlorobenzenes, hexachlorobutadiene, bis(2-ethylhexyl) phthalate and polynuclear aromatic hydrocarbons (PAHs). Concentrations of hexachlorobenzene and hexachlorobutadiene have been found to be higher than in any other Great Lakes region.

Sediment samples taken in the old freighter slip, near the mouth of Strong Brook, had extremely elevated concentrations of oil and grease (33,000 mg/kg), lead (350 mg/kg) and zinc (830 mg/kg), indicating a pollutant source from Strong Brook (Environmental Research Group 1979). Contaminant levels in the river decrease downstream from the Fields Brook confluence and preliminary sampling, using sediment cores, indicates that the underlying sediment may be more polluted than surface sediments.

Although sediments in the outer harbor are classified as nonpolluted to moderately polluted, a comparison of 1984 and 1988 data reveals that PCB concentrations are increasing, indicating that polluted sediments are continuing to move downstream. Any further increases in contaminant concentrations could result in reclassification of the sediments to heavily polluted, thereby requiring confined disposal.

The most comprehensive sediment sampling effort on Fields Brook was conducted as part of the Superfund Remedial Investigation Report (CH$_2$M Hill 1985). Many organic chemical compounds were quantified, including methylene chloride, 1,1,2,2-tetrachloroethane, tetrachloroethene, trichloroethene, 1,2 transdichloroethene, PAHs, hexachlorobutadiene, phthalates, PCBs, chlorobenzene, hexachlorobenzene and other chlorinated benzenes. Mercury was above background levels at almost every sampling site and elevated levels of arsenic, zinc, chromium, lead and titanium were also found. The highest concentrations of pollutants appeared to be localized in the mid-segment of Fields Brook, adjacent to the concentrated industrial complex. Although sources of pollutants can be inferred from existing data, further studies are being conducted under the Superfund Investigation Program to substantiate assumptions and identify any active sources.

Exploratory studies by Veith et al. (1981) of whole body, multi-species fish composites from the Ashtabula River-Fields Brook area revealed extremely complex mixtures of bioaccumulated chemicals. At least 45 chemicals were identified including high concentrations of hexachlorobutadiene, hexachlorobenzene, octachlorostyrene, PCBs and other chlorinated compounds. Subsequent analyses of single species whole body and fillet samples collected between the mouth of the Ashtabula River and Fields Brook confirmed high concentrations of PCBs, hexachlorobenzene, pentachlorobenzene and tetrachloroethane, as presented in Table 8 (Armstrong 1983). This information formed the basis for issuing a fish consumption advisory by the Ohio Department of Health and Ohio Environmental Protection Agency (EPA) in 1983 (Ohio Department of Health and Ohio EPA, 1983). The advisory recommends that no fish caught in the lower 3.2 km (2 miles) of the river and harbor be eaten, citing PCB concentrations ranging from 2.4 to 58.3 mg/kg in the edible portion. The current U.S. Food and Drug Administration action level for PCBs is 2 mg/kg for the edible portion of fish.

A recent biological survey conducted by Ohio EPA documented a diverse fish community upstream from the Area of Concern (Ohio EPA 1990). This area remains largely in its natural state and includes good pool and riffle habitat. The diversity and numbers of fish decrease downstream from Fields Brook in the river navigation channel and show continued impact toward the river mouth; the poorest communities were located near the vertical bulkheads of the commercial shipping docks. Results from the Ashtabula River survey were compared with results from a survey on Conneaut Creek, a tributary located just east of Ashtabula, with a similar river and harbor hydrography but without the industrial impact. Biological index values for the Ashtabula River were similar to those from Conneaut Creek. Based on these data and information on aquatic habitat, it was hypothesized that habitat destruction due to river development in the Ashtabula River had as great an impact on the

TABLE 8. Mean Concentrations of Selected Organic Compounds in Fish Samples Collected within a 0.8 km (0.5 mile) Area of the Fields Brook–Ashtabula River Confluence in 1980 and 1981

Samples	Length (mm)	Weight (g)	Lipid Content (mg/g)	Total PCB (mg/kg)	Hexachlorobenzene (mg/kg)	Pentachlorobenzene (mg/kg)	1,1,1,2-Tetrachloroethane (mg/kg)	1,1,2,2-Tetrachloroethane (mg/kg)
Carp, whole (25 samples)								
Range	172–454	184–2910	51–304	4.0–152.6	0.28–7.49	0.06–0.63	1.08–5.70	0.01–0.62
Mean	318	1053	211	50.0	3.40	0.22	2.60	0.22
Std. Dev.	77	701	234	46.1	2.40	0.16	1.80	0.15
Carp, fillet (21 samples)								
Range	75–506	246–4320	21–267	2.4–58.3	0.37–5.80	0.03–0.36	0.02–3.66	0.02–0.51
Mean	359	2317	203	19.0	2.0	0.14	1.50	0.16
Std. Dev.	107	2841	250	18.0	1.5	0.08	1.00	0.11
Bullhead, whole (11 samples)								
Range	146–282	73–574	25–79	0.94–8.53	0.04–1.13	0.01–0.06	0.01–0.39	0.004–0.04
Mean	197	222	77	4.44	0.57	0.03	0.07	0.02
Std. Dev.	47	175	62	2.86	0.37	0.02	0.13	0.01
Bullhead, fillet (26 samples)								
Range	ND	39–574	4–138	0.60–28.0	0.003–4.66	0*–0.19	0*–0.84	0.002–1.05
Mean	ND	195	38	5.6	0.96	0.04	0.06	0.07
Std. Dev.	ND	152	28	6.4	1.11	0.04	0.17	0.20

Note: The fish consumption advisory was largely based on this information (Armstrong 1983). 0* indicates less than detection limit; ND = not determined.

fish community as the chemical pollution. This conclusion was reinforced by the discovery of a diverse, highly productive community in a well-vegetated area inside the inner breakwall in the Ashtabula Harbor (Ohio EPA 1990).

During this survey, a population of brown bullhead afflicted with numerous lip and skin tumors was discovered inside the west breakwall of Ashtabula Harbor. Similar anomalies were not found in any fish further upstream or near the mouth of Fields Brook. These observations suggest a significant pollution source other than Fields Brook near the mouth of the river. Coal dust from the adjacent coal yard and conveyor may be a source of PAHs, which could be the cause of the tumors. Yet, Conneaut Harbor has a coal yard and tumors have not been found on fish from this area. Differences in the two facilities may account for these discrepancies; Conneaut facility is located downwind from the river and does not have a conveyor. Further investigations are underway.

A summary of the 14 potential beneficial use impairments listed in Annex 2 of the Great Lakes Water Quality Agreement and their significance to the Ashtabula River Area of Concern is presented in Table 9. All major problems in the Ashtabula River Area of Concern are directly related to contaminated sediments.

A number of remedial activities were initiated under Superfund (Comprehensive Environmental Response, Compensation, and Liability Act) for the Ashtabula River Area of Concern before the advent of the RAP process, but numerous obstacles have prevented implementation. A Remedial Investigation Report completed in 1985 assessed the seriousness of the contamination in Fields Brook and served as the basis for evaluation of remedial alternatives for the site (CH$_2$M Hill 1985). A Record of Decision was signed in 1986 that selected remedial options and outlined additional studies needed before cleanup could be implemented; however, several problems delayed further activities, including waiting for reauthorization of Superfund monies and extended, complicated negotiations between U.S. EPA and the potentially responsible parties.

The Buffalo District of the U.S. Army Corps of Engineers (Corps) has attempted to dredge the Ashtabula River since the early 1970s but has been frustrated by the lack of an acceptable disposal site. Even before the sediments were classified as toxic, the Corps attempted to site a confined disposal facility (CDF) in Ashtabula un-

TABLE 9. Summary of Potential Use Impairments as Cited in Annex 2 of the Great Lakes Water Quality Agreement and Their Significance to the Ashtabula River Area of Concern

Use Impairment	Significance to Ashtabula Area of Concern
1. Restrictions on fish or wildlife consumption	Fish consumption advisory for all species due to high concentrations of PCBs, hexachlorobenzene, pentachlorobenzene and tetrachloroethane.
2. Tainting of fish and wildlife flavor	None reported.
3. Degraded fish and wildlife populations	Fish community status is fair to poor. Affected by chemical pollution immediately downstream from Fields Brook, but main impact is from habitat destruction due to marina development and commercial dock facilities.
4. Fish tumors or other deformities	Large incidence of skin tumors in brown bullhead population inside west breakwall.
5. Bird or animal deformities or reproductive problems	None reported.
6. Degradation of benthos	Recent study not yet summarized. Past studies indicate dominance of pollution tolerant species in lower river.
7. Restrictions on dredging	Navigation channel has not been dredged since 1964 due to lack of a disposal site for heavily polluted and toxic sediments.
8. Eutrophication or undesirable algae	No algal blooms. Available data are ten years old, but no dissolved oxygen depletion or phosphorus problems were reported.
9. Restrictions on drinking water consumption or taste and odor problems	None reported, but possible chemical contamination is being investigated under Superfund.
10. Beach closings	None reported.
11. Degradation of aesthetics	Debris and muddy water after rainstorms.
12. Added costs to agriculture or industry	None reported.
13. Degradation of plankton populations	Unknown. Bioassays show sediment to be toxic to *Ceriodaphnia*.
14. Loss of fish and wildlife habitat	Marina construction, vertical bulkheads and depth associated with commercial loading docks have severely impacted fish community.

der the authority of U.S. Public Law 91–611. This law provided for the construction of CDFs for polluted materials dredged from the Great Lakes and was the funding source for all CDFs constructed along the southern shore of Lake Erie. A CDF was never built in Ashtabula because of disagreement on site location and construction costs and therefore the river channel has not been dredged since 1962.

Sediment sampling data from 1979 and 1983 confirmed that sediments were highly polluted and toxic due to PCB concentrations exceeding 50 mg/kg, further complicating the dredging and disposal issue. The Corps submitted an Environmental Impact Statement in 1983 proposing an extensive one-time dredging project to remove and dispose all contaminated sediment in the federal navigation channel by removing an additional 0.6 m (two feet) below the original project depth of 4.9 m (16 feet) below low water datum.

Several obstacles impeded the dredging project:

1. If the river was dredged before remedial activities on Fields Brook were completed, would the river channel become recontaminated?
2. Would dredging uncover more severely contaminated sediment?
3. Who would be responsible for cleanup of contaminated sediments outside the federal channel?
4. Could heavily polluted and toxic sediments be separated for disposal?
5. What permits and regulations must be followed to dispose of toxic sediments?
6. Where could a suitable disposal site be located?
7. Who would pay for the project?
8. What precautions would be taken to minimize resuspension of the contaminated sediments in the water during the dredging and dewatering activities?

The Corps evaluated over 20 sites as possible locations for a confined disposal facility, but most were rejected as environmentally unacceptable. Under the authority of U.S. Public Law 91-611, the local government was responsible for providing a disposal site and obtaining all necessary permits or no action would be taken. A site could not be agreed upon by the local jurisdictions and the Corps dropped their support of the proposed project in May 1983.

In 1987, the Buffalo District of the Corps submitted another proposal recommending disposal of all materials (152,920 m³ toxic sediment and 229,380 m³ highly polluted sediment) at an upland, privately owned site. Finally, this plan was agreeable to all parties and the Corps had $11 million in their budget to implement the project. Escalating costs for handling toxic sediments and the need for disposal liability insurance resulted in a revised estimate of $25 million for the project. The Washington Office of the Chief Engineers (OCE) was unwilling to commit that much money to one project and stated that it was not their responsibility to clean up toxic sediments. The OCE's present policy is to not construct any new CDFs under U.S. Public Law 91–611, so funding for the Ashtabula project would be taken from their harbor maintenance budget. A 25 percent budget cut under the Reagan Administration forced the OCE to focus their attention on commercial navigation rather than environmental cleanup and recreational navigation channels, as is the case with the Ashtabula River.

The RAP Process

The Ohio EPA RAP program was initially supported by a grant from U.S. EPA's Great Lakes National Program Office (GLNPO). A GLNPO-contracted consultant, Science Applications International Corporation, wrote the original background report in 1986. As the process continued, different funding sources were sought to pay for the development of each Ohio RAP; funding for the Ashtabula River RAP is provided by a grant from GLNPO, state general revenue funds and other federal funds allotted to state agencies in support of programs under the U.S. Clean Water Act (Sections 106, 205j).

After the original background report was completed, Ohio EPA held a public meeting in October 1987 to discuss this report and explain the RAP process. The extent of environmental contamination, the length of time this contamination had existed, and the lack of actual cleanup actions had helped to create a cynical local attitude. This atmosphere prevailed at the first meeting and was the first obstacle Ohio EPA had to overcome. However, the emphasis on public involvement at all stages of the RAP process convinced several people that the RAP would not be just another study. In addition, Ohio EPA and Ohio State Senator Robert Boggs invited community members to another meeting in February 1988 to discuss the environmental problems of the Ashtabula Area of Concern, the RAP guidelines, and a plan to establish local input at the early stages. Senator Boggs is a highly-respected local politician and his invitation influenced the active participation of many local citizens and community leaders.

As a result of the February meeting, the Ashtabula River RAP Public Advisory Council was established (Ohio EPA 1989). Membership on the council is voluntary but all pertinent groups are represented, including local government, industry, business, environmental groups, state regulatory agencies, unaffiliated citizens and special interest groups. U.S. Representative Dennis Eckart and his staff are active in the RAP program and serve as ex-officio members of the council. Two committees were formed to address specific technical and communications issues. These committees decided to delegate all RAP writing responsibilities to Ohio EPA but remain active in setting goals, reviewing drafts, supporting public information activities and seeking solutions and funding.

Due to the complicated technical and legal issues associated with the Superfund cleanup on Fields Brook and the fact that the remedial alternatives and costs are addressed under Superfund, the RAP council focused its efforts on the Ashtabula River cleanup. Fields Brook activities are being followed closely and some public involvement activities required by Superfund may be combined or coordinated with those of the RAP.

The RAP Advisory Council is a relatively small group compared to those in other Areas of Concern around the Great Lakes, but it represents a population base of about 30,000 residents. The council has been very effective in attaining local government suport and council meetings serve as forums to answer questions and explain state and federal regulatory programs. They also provide the opportunity for frank discussion, presentation of varying perspectives, and coordination toward a common goal.

Ohio EPA views the RAP process as a step toward achieving the objectives of the Great Lakes Water Quality Agreement, with the focus on restoration of the 14 beneficial uses for the Area of Concern (Ohio EPA 1989). Stage 1 of the RAP is focussing on problem definition and description of causes, while Stage 2 will select and implement those remedial actions needed to restore these uses; considerable coordination will be required to reach some difficult decisions. Stage 2 also will require more of a team approach than used in Stage 1, but the communication network assembled during the preparation of Stage 1 will help to facilitate development of the Stage 2 RAP.

Accomplishments of the Ashtabula River RAP Program

For a RAP program to be truly successful, it must highlight remedial actions completed and milestones reached in the area, regardless of whether or not they were a direct result of the RAP. This serves to track overall progress and also to keep enthusiasm high. A number of activities had already been initiated in the Ashtabula River Area of Concern under other programs, but the RAP provided more opportunities for public involvement, education and it served as a catalyst to accelerate stalled projects.

Perhaps one of the greatest impacts of the RAP is the revitalized community cooperation. Local residents have realized that the RAP may be their last chance to restore the river and have rallied to do everything they can to secure funding and implement cleanup. As was noted earlier, community indecision in the past resulted in the cancellation of two river dredging plans.

When the Office of the Chief Engineers decided not to fund the second proposal for river dredging, the RAP Council organized a letter writing campaign to obtain federal support. A detailed slide program and accompanying report were prepared to emphasize the immediate need for river dredging, and Congressman Eckart and council members met with the Corps of Engineers in Washington, D.C., to present the prepared slide program. Other attendees at the meeting included the chief administrators and key personnel from U.S. EPA Region V, Ohio EPA, Ohio State Senator Robert Boggs, representatives of the Ohio Department of Natural Resources and Ohio's U.S. Senators John Glenn and Howard Metzenbaum. Although the Office of the Chief Engineers did not agree to fund the total operation, they did request additional information for further consideration of a scaled-down project. The local RAP contingency felt they scored a victory by getting all the deciding authorities together in one place to collectively discuss a solution. In further negotiations with the Corps, Congressman Eckart obtained a $300,000 commitment to dredge at least one meter (3.3 feet) in the channel to alleviate a navigational hazard. This could be done in the near future if it doesn't involve handling toxic sediments and a disposal site can be found.

The Corps wasn't the only group that required additional information. The Superfund Record of Decision for Fields Brook specified that an intensive study of the river be completed to determine the extent of contamination and additional sources, if any. Five potentially responsible parties agreed to fund the river study and suggested that their sampling effort be combined with that of the Corps to obtain a more complete data set and avoid duplication of effort. Coordinated planning of this study delayed the sampling for eight months due to the extensive quality assurance/quality control review process stipulated under Superfund, but this study will provide all additional information needed to dredge the river. Preliminary results indicate the top meter of sediment is not considered toxic, and it is highly probable that some navigational dredging will occur in 1991.

Realizing that funding for the river dredging would not be provided solely by the federal government, Ohio EPA and the RAP Council began to investigate other Ohio resources. After discussion with a number of state agencies, they concluded that state funding would best be provided from one source rather than trying to coordinate small contributions from many agencies. Ohio House Bill 592, Ohio's solid waste law, created a fund from tipping fees to be used for cleanup of state hazardous waste sites not on the National Priority List. Ohio committed $7 million of that fund toward Ashtabula River cleanup, $3.5 million for 1989–1990 and $3.5 million for 1991–1992. This was the first money from this fund to be designated for a specific site and Congressman Eckart is using this commitment as leverage to obtain more federal funding. It is extremely rare for a state to guarantee financing before federal funding is approved, and it emphasizes the high priority the State of Ohio has placed on this project.

At the local level, the Ashtabula City Council passed a law to tax boat docks and launching ramps on the river. The estimated annual revenue of $30,000 will be earmarked for river improvements such as pump-out facilities, lighting and shoreline improvements. Although not associated with dredging, these actions would enhance use of the river.

Findings and orders issued in 1989 under the National Pollutant Discharge Elimination System permit for Occidental Chemical Company, one of the settling potentially responsible parties on Fields Brook, required the company to donate $7,500 to the RAP program. The donation was in lieu of a fine for not meeting compliance with permit limitations. Occidental's problem was not in its process water discharge, but rather in runoff from onsite contamination resulting from past disposal practices. Contaminants include volatile organic compounds, particularly carbon tetrachloride and chloroform. While Occidental was in the process of constructing retaining walls, a series of drains and a collection system to contain and treat the runoff, the schedule for completion did not meet the compliance deadline. This system has recently come online and is expected to control the problem.

A RAP newsletter is produced quarterly by Ohio EPA and the RAP Council to keep the local community informed of issues related to river cleanup. A brochure and several fact sheets also have been prepared and are used to promote the RAP process at area sport shows, festivals and special events. RMI, another settling party on Fields Brook, has produced several videos highlighting the river's problems and the need for dredging in association with the RAP Council. Council members have presented them at numerous meetings and events, which receive frequent media coverage. Council members have participated in Great Lakes United workshops on public involvement and have shared ideas with representatives from other Areas of Concern, and the RAP program has helped to educate the public concerning the environmental problems of the Ashtabula River as well as the state and federal regulatory programs. RAP Council members can now be a part of finding solutions, rather than just complaining to local, state, and federal governments that *they* should do something.

As previously mentioned, most of the technical decisions required for Fields Brook cleanup will continue to be made under the Superfund program. A Natural Resource Damage Assessment investigation also is underway under Superfund to determine if the state is justified in submitting a claim for additional compensation for damage to the natural resources of the area.

On a broader scale, the RAP process is beginning to change the way Ohio EPA operates. A Lake Erie Group has been created to deal with lake issues, and particularly RAPs. An inhouse Lake Erie Oversight Committee, comprised of representatives from each division, meets monthly to discuss Lake Erie issues and encourage information exchange. The RAP multi-media concept is being considered for application at other sites located outside the Lake Erie drainage basin.

Future Challenges

Although significant progress has been achieved, the biggest problems have yet to be solved. Responsibility for the Fields Brook cleanup lies with U.S. EPA and the potentially responsible parties, but responsibility for cleanup of the Ashtabula River is still undecided.

A number of other questions relevant to river cleanup must be addressed. Sediments in one section of the river are classified as toxic, while the rest of the sediments are heavily polluted, and disposal requirements are different for each. Can the sediments be adequately separated? Where will the sediments be disposed? Who will be liable for the toxic sediments after they are placed in a Toxic Substances Control Act landfill? Can the disposal of toxic sediments be combined with the disposal of similar sediments from Fields Brook? Would small-scale removal of surface sediment, to alleviate the navigation hazard, uncover more severely contaminated sediment? What precautions must be taken to minimize resuspension of polluted sediments during dredging and dewatering activities? How will Fields Brook dredging impact the river if it is done after the river is dredged? Who will fund the river dredging project? If the cleanup operation is a joint funding and implementation effort, who will be responsible for coordination? All these questions remain to be answered.

The RMI Extrusion Plant on Fields Brook has emitted radioactive materials in the past, particularly uranium, and some elevated concentrations of radionuclides have been found onsite. These contaminated areas are being cleaned up, but there is some public concern that Fields Brook and the river may contain high levels of radionuclides from past discharge and air emissions. No evidence of any contamination other than that found onsite has been found, and further tests of the river sediment have been completed to alleviate this public concern. One option for disposal of heavily polluted material dredged from the navigation channel in 1991 is a confined disposal facility in Cleveland. Without information on radionuclides, however, public opposition may prevent use of this facility and no other sites are immediately available.

The discovery of tumors in brown bullhead at the mouth of the river creates another series of questions to be answered: what is causing the tumors? Are they benign or malignant? Can humans be affected by contact with sediment or water in this area? Are there further undiscovered sources of pollutants?

The impact of atmospheric pollutant deposition in the Area of Concern is unknown. Local air emissions data are available, but how do they relate to contamination in the river? The current environmental quality of the Lake Erie nearshore area also is unknown. The most recent chemical data are more than 10 years old. A search of file information indicates that a mercury contamination problem may be associated with one of the Lake Erie dischargers; the nearshore zone needs further investigation.

Another challenge for the cleanup program is to attain a balance between economic growth and environmental quality. Industry is growing in Ashtabula and several companies are planning extensive expansions. The local government is demanding a guarantee that the environment will be protected before approving industry expansion through tax abatements. Today's more restrictive National Pollutant Discharge Elimination System permits will require state-of-the-art treatment facilities for new plants and specific biological criteria that must be met. Several industries on Fields Brook discharge high concentrations of total dissolved solids, frequently in excess of their permit limits. Investigations are underway to determine if these concentrations are toxic to biological communities. Can Fields Brook even be considered a suitable habitat for supporting a viable warmwater fishery if its flow consists mostly of effluent discharge?

Despite the contamination problems, Ashtabula Harbor remains a popular recreation spot. Marinas are still expanding, bringing more boat traffic into the area. Fish and wildlife habitat already has been severely disturbed. How can further habitat loss be prevented and lost habitat restored while still expanding the recreation industry?

From a policy perspective, implementing a multi-media ecosystem approach to restore beneficial uses in the Areas of Concern is much more difficult than managing one specific program. It will require considerable coordination within Ohio EPA as well as with other state and federal agencies. The needs of the local community also must be considered in any decisions.

Perhaps the biggest challenge for all Areas of Concern is funding. Significant federal legislation will be needed to authorize these funds and priorities must be set to determine which areas will receive the funds. States will have to balance their priorities between the Great Lakes and inland sites. Certain sites in Ohio, for example, are significantly more contaminated than any of the Areas of Concern (i.e. Mill Creek, Mahoning River). Where will the funds come from for those cleanups and should they have a higher priority? These are questions that must still be addressed.

Summary

The problems of the Ashtabula River Area of Concern began nearly 50 years ago. As the chemical industry thrived and expanded, Fields Brook and the Ashtabula River became increasingly contaminated with metals and numerous organic chemical compounds. Sediments in the river's navigation channel have not been dredged since 1962 due to lack of a suitable disposal site for the contaminated sediments. A fish consumption advisory for all species exists in the Area of Concern, and the recent discovery of tumors in brown bullhead raises the question of whether or not to expand the advisory to include human contact.

Remedial actions implemented under the U.S. Clean Water Act, U.S. Resource Conservation and Recovery Act and other legislation have stopped or slowed the flow of pollution, but a tremendous cleanup operation lies ahead. The RAP program provides a means for the local community to communicate directly with state and federal government agencies and the chance to help restore the environmental integrity of the area. In the Ashtabula River Area of Concern, the RAP program is working.

REFERENCES

Aquatech. 1983. *Analysis of sediment from Ashtabula River, Ashtabula, Ohio.* Prepared for the U.S. Army Corps of Engineers #DACW 49-82-C-0062.

Armstrong, D.E. 1983. *Preliminary tabulation of data for selected organic compounds in fish samples collected from the Fields Brook–Ashtabula River area in 1980 and 1981.* Memo in file at Ohio EPA.

CH$_2$M Hill. 1985. *Final remedial investigation report, Fields Brook site, Ashtabula, Ohio.* Prepared for U.S. EPA, Hazardous Site Control Division, 68–01–6692.

Environmental Research Group, Inc. 1979. *Field sampling analysis of core sediment samples, Ashtabula River, Ohio.* Columbus, Ohio.

Ohio Department of Health and Ohio EPA. 1983. *Joint statement of health advisory for consumption of Ashtabula River fish.* Columbus, Ohio

Ohio EPA. 1989. *Ashtabula River RAP. Draft Stage 1.* Ohio EPA, Division Water Quality Planning and Assessment. Columbus, Ohio.

Ohio EPA. 1990. *Ashtabula River Natural Resource Damage Assessment. Biological Communities, Ashtabula County, Ohio.* Ohio EPA, Division of Water Quality Planning and Assessment. Columbus, Ohio.

U.S. EPA. 1977. *Guidelines for the pollutional classification of Great Lakes Harbor sediments.* U.S. EPA, Region V, Chicago, IL.

Veith, G.D., D.W. Kuehl, E.N. Leonard,K. Welch, and G. Pratt. 1981. Polychlorinated biphenyls and other organic residues in fish from major United States watersheds near the Great Lakes, 1978. *Pesticides Monitoring Journal* 15(1):1–8.

Chapter 7

The Quest for Clean Water: The Milwaukee Estuary Remedial Action Plan

Dan Kaemmerer, Audrey O'Brien, Tom Sheffy, and Steve Skavroneck

> "Clean water is vital to the future prosperity of the Milwaukee area and the entire state. The Milwaukee Estuary Remedial Action Plan will help ensure that prosperity by cleaning up the estuary and keeping it clean."
> John Norquist
> Mayor of Milwaukee

Introduction

Water quality problems in the Milwaukee area represent the legacy of decades of use and abuse. These problems were never anticipated, yet through ignorance and negligence have manifested themselves as contaminated sediments, contaminated fish and wildlife, declining fish and wildlife populations, poor water quality, habitat loss and potential health threats to the residents of this region. An unprecedented opportunity exists to solve these problems and dramatically improve the quality of water resources in the Greater Milwaukee area. The Milwaukee Estuary remedial action plan (RAP) is the blueprint for fulfilling this opportunity. From the perspective of the Wisconsin Department of Natural Resources, the City of Milwaukee, Milwaukee County, the Milwaukee Metropolitan Sewerage District and the numerous business leaders, organizations and citizens involved in preparing this document, the RAP symbolizes everyone's readiness and commitment to restore the environment in order to sustain all desired uses of Milwaukee's waterways. This chapter documents the process being used to develop the Milwaukee Estuary RAP and ensure broad-based community support, and highlights the successes and challenges of the Milwaukee RAP process.

Problems and Use Impairments in the
Milwaukee Estuary

The Milwaukee Area of Concern includes the Milwaukee Harbor, the Milwaukee River downstream of the North Avenue Dam, the Menomonee River downstream of 35th Street, the Kinnickinnic River downstream of Chase Avenue and nearshore areas of Lake Michigan (Figure 12). While problems and use impairments are summarized in Table 10, more precise description of each of the impaired beneficial uses and causes is presented below:

Description of Identified Impaired Uses in the Milwaukee
Area of Concern

Restrictions on Fish and Wildlife Consumption
Fish. Wisconsin's fish consumption advisory recommends that people avoid the consumption of crappie, northern pike, carp, redhorse, smallmouth bass, catfish, white sucker or larger lake trout (over 58 cm or 23 inches), chinook salmon (81 cm (32 inches) or greater), or brown trout (over 58 cm or 23 inches) taken from the Milwaukee Estuary. The advisory further recommends that women and children avoid the consumption of medium-size lake trout (51 cm to 58 cm or 20 to 23 inches), coho salmon (over 66 cm or 26 inches), chinook salmon (53 cm to 81 cm or 21 to 32 inches), and brown trout over 58 cm (23 inches). Perch, smelt, pink salmon, rainbow trout, brook trout, and smaller lake trout, coho salmon, and chinook salmon pose the lowest health risk.

PCB contamination is the principal reason for these advisories. The pesticides chlordane and dieldrin also have been found to exceed consumption advisory levels in lake trout, while lindane, dieldrin, DDE, mirex, dioxins and furans have been found in northern pike, carp, and smallmouth bass.

Wildlife. A similar consumption advisory exists for waterfowl caught in the vicinity of the Milwaukee Harbor. No one should eat mallard, black ducks, scaup, or ruddy ducks. Contaminated waterfowl could be a regional problem that may not be directly attributable to pollution from this Area of Concern. Again, PCB is the principal contaminant. Mercury and the pesticides dieldrin, DDE, DDD and DDT also have been found in wildlife.

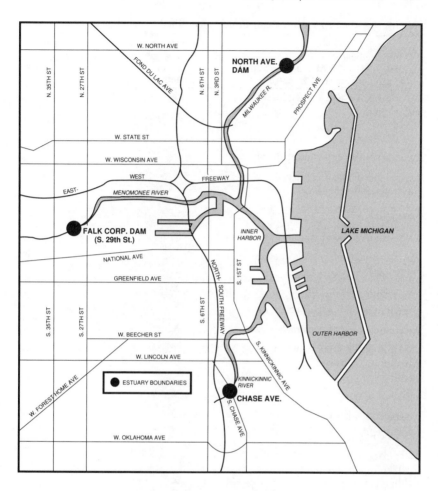

Fig. 12. The Milwaukee Estuary Area of Concern as defined by the Milwaukee River downstream of the North Avenue Dam, the Menomonee River downstream of 35th Street, the Kinnickinnic River downstream of Chase Avenue, and the nearshore areas of Lake Michigan

Degraded Fish Populations

Fish populations in the Area of Concern are degraded due to poor water quality. While 35 species have been identified in the Milwaukee Estuary, many are transient, nonresident species. Resident species in the estuary are the more pollution tolerant species such as carp. In addition, periodic fish kills occur due to low dissolved oxy-

TABLE 10. Problems and Use Impairments Identified in the Milwaukee Estuary

IJC Impaired Uses (if identified in the Area of Concern, these problems need to be corrected)	Impaired Use in the Estuary (caused by water quality problems)	Use is unimpaired (not caused by water quality problems or toxic chemicals)
Restrictions on fish and wildlife consumption		
Fish	X	
Wildlife	X	
Tainting of fish and wildlife flavor		X
Degraded fish and wildlife populations (diversity and abundance)		
Fish	X	
Wildlife		X[a]
Fish tumors or other deformities	X[b]	
Bird or animal deformities or reproductive problems		X[a]
Degradation of benthos	X	
Restrictions on dredging activities	X	
Eutrophication or undesirable algae	X	
Restrictions on drinking water consumption or taste and odor problems		X
Beach closings	X	
Degraded aesthetics	X	
Added costs to industry		X
Degradation of phytoplankton and zooplankton populations		
Phytoplankton	X	
Zooplankton	X	
Loss of fish and wildlife habitat		
Fish habitat	X	
Wildlife habitat		X[a]

[a]These use impairments are not caused by water quality problems in the Area of Concern; however, they may be impacted by poor physical habitat caused by urbanization. The Area of Concern is completely urban, serving residential, commercial and industrial uses. Improvement of habitat may restore some of these uses. Problems caused by poor quality habitat or lack of habitat are discussed in this chapter. In addition, goals and objectives address problems resulting from poor physical habitat.

[b]Available data, when compared to other locations with known fish tumors, indicates that sediment PAH concentrations are sufficient in some portions of the estuary to induce fish tumors. However, because of the age of this data base, further research may be warranted.

gen levels (caused by nutrient enrichment, high levels of suspended solids, and high water temperatures). In October 1989, a fish kill of 100,000 catfish, walleye, northern pike, gizzard shad and other species occurred in the Burnham Canal off the Menomonee River when high concentrations of heavy metals were found in the water column. Another fish kill occurred in the South Canal off the Menomonee River in January 1990. Major fish kills also were reported in these canals in February 1979, and August 1984.

The lack of species diversity is due to poor water quality (high turbidity, low dissolved oxygen levels, hypereutrophic conditions) caused by nonpoint sources of pollutants upstream and from combined sewer overflows, erosion, and other nonpoint runoff from adjacent land. Poor physical habitat such as shoreline steel pilings, concrete river channels, and dams contribute to degradation of fish populations.

Fish Tumors or Other Deformities
There is no evidence of fish tumors or other deformities in the estuary. However, when data from the Milwaukee estuary are compared with data from Great Lakes locations where there is evidence of fish tumors, it is clear that sediment PAH levels in the Inner Harbor and the Kinnickinnic, Menomonee and Milwaukee Rivers are sufficient to cause elevated incidence of cancer in susceptible species of demersal fish. Average concentrations of fluoranthene and pyrene are about twice as high in the Milwaukee River and its tributaries as in locations where demersal fish are known to have high cancer rates. In a few sections of the estuary, PAH concentrations either approach or breach the upper range value for these carcinogens. While fish tumors or other deformities are presumed to be present, further research to substantiate the existence of this problem in the estuary is needed.

Degradation of Benthos
Species diversity is poor in the Area of Concern. Dominant species are very pollution tolerant (*Asellus intermedius; oligochaeta*) and indicate organically enriched and silty substrate. Causes for this impairment include sedimentation from upstream, local erosion and runoff, and the pollutants in sediments and water.

Restrictions on Dredging Activities
Moderate to heavily polluted dredge spoils (based on U.S. EPA's 1977 Guidance) are located in the confined disposal facility immediately north of South Shore Park, which is expected to be filled to capacity in 1993. Open water disposal is prohibited. Sediments in the outer harbor and the Kinnickinnic, Menomonee and Milwaukee Rivers are moderately to highly polluted with PCBs, COD, phosphorus, ammonia nitrogen, lead, zinc, cadmium, arsenic, copper, dioxins, furans, PAHs, oil, and grease. Sources of these contaminants include point source discharges, storm sewer runoff, upstream nonpoint sources, storage pile drainage, spills and combined sewer overflows.

Eutrophication and Undesirable Algae
Eutrophication in the area is a result of excessive nutrient loadings from combined sewer overflows, sanitary sewer overflows, upstream nonpoint pollution, storm sewer discharge, urban runoff, and other sources. These nutrient loadings have caused fish kills, reduced water quality and clarity, and induced the excessive growth of periphyton on shoreline and lake front structures.

Beach Closings
Beaches in the nearshore areas of Lake Michigan outside the outer harbor are closed for 48 to 96 hours when rainfall levels are greater than 0.76 cm (0.3 inches). Restrictions on full or partial body contact are suggested for the rivers and outer harbor due to high levels of fecal coliform bacteria from combined sewer overflows, sanitary sewer overflows and urban and rural nonpoint pollution.

Degradation of Aesthetics
After storms, considerable debris and garbage accumulate in the Milwaukee estuary, and litter and debris are regularly visible. A "skimmer" operated by the Milwaukee Metropolitan Sewerage District removes debris from the surface water of the rivers on a regular basis throughout the summer. Combined sewer overflows and carelessness along the shorelines contribute litter and debris to the water. Unfortunately, many area residents do not have a "stewardship" or "caretaker" ethic toward the harbor area, which is a major obstacle to overcome.

Loss of Fish Habitat

Macrophytes found in the area are pollution tolerant or nonexistent. Sediments are contaminated, providing an avenue for bioaccumulation, and continual nutrient and sediment loading further degrades available habitat. The substrate is unable to support a balanced ecosystem due to concrete channelization in several upstream reaches of the rivers.

River habitat has been degraded due to combined sewer overflows, sanitary sewer overflows, erosion from upstream and local sources, runoff from adjacent land and urbanization. Poor physical habitat such as shoreline steel pilings, concrete river channels and dams also impede fish migration and spawning.

Degradation of Phytoplankton and
Zooplankton Populations

Phytoplankton populations are impaired in the Milwaukee Area of Concern due to poor water quality. The incidence of brackish water species indicates the community could be influenced by chlorides. The total number of phytoplankton cells/milliliter is higher inside the harbor than at 1.6 km (one mile) outside the breakwall. The greater concentrations of nutrients within the harbor not only allow the more pollution tolerant organisms to gain a competitive advantage over other organisms, but also add to the quantity of organisms found.

The eutrophic condition of the rivers discharging into the harbor contributes to a quantity-rich, species-poor community. The predominance of periphytic species of diatoms in the outer harbor, various spectral analyses and water chemistry data indicate that the three rivers have a significant influence on the phytoplankton community of the outer harbor. Likewise, zooplankton populations are also affected by the poor water quality conditions of the rivers and harbor. Data collected to date by the Milwaukee Metropolitan Sewerage District indicate fewer species present in the outer harbor; the dominant species found is the *Bosmina spp.*, a type of zooplankton tolerant of eutrophic conditions. Like the phytoplankton, zooplankton communities are species-poor, quantity-rich communities that thrive in the eutrophic conditions of the rivers and harbor. An exotic species, the spiny water flea (*Bythotrephes cederstroemi*), has been found in the Milwaukee harbor area and may affect the food chain by preying on other zooplankton species that are valuable as fish food. The phytoplankton and zooplankton populations also are affected by the physical barrier of the breakwall, which impedes exchange with Lake Michigan.

Nonpoint and point source pollution cause poor water quality conditions such as increased temperatures, increased chloride levels, excess nutrient and sediment loadings, and reduced light penetration.

Uses Not Documented to Be Impaired Because of Water Quality Problems in the Milwaukee Area of Concern

Tainting of Fish and Wildlife Flavor
There is no evidence of a tainting or flavor problem in fish or wildlife.

Bird or Animal Deformities or Reproductive Problems
No bird or animal deformities have been documented in the Area of Concern, and no evidence exists that levels of contaminants are causing deformities or reproductive problems in birds or animals in the estuary.

Restrictions on Drinking Water Consumption or Taste and Odor Problems
Milwaukee's water treatment facilities are meeting all drinking water primary (health) and secondary (taste and odor) standards.

Added Costs to Agriculture or Industry
In a survey by the Metropolitan Milwaukee Association of Commerce, two industries indicated they must treat water drawn from the estuary prior to using it for industrial processing (e.g. cooling). Wisconsin Electric Power Company draws water from the Menomonee River and chlorinates it prior to use. This treatment is a standard procedure to guard against biofouling equipment and is implemented regardless of water quality. Similarly, Eaton Corporation provides water softening treatment to water drawn from the Milwaukee Harbor. Although this type of treatment does not justify documenting this as an impaired use, the water quality problems that make treatment necessary will be investigated as part of the RAP.

*Degraded Wildlife Populations (Diversity
and Abundance)*

While this is not an impaired use as a result of poor water quality conditions, lack of a suitable habitat has resulted in a poor diversity and limited numbers of wildlife species in the Milwaukee Estuary. No wetlands exist and little habitat is available on streambanks. Like other parts of the Great Lakes region, contaminants are bioaccumulating in waterfowl and other species. In the Milwaukee estuary, a snapping turtle caught in 1983 contained 0.2 μg/kg of dioxin and 0.1 μg/kg of furans; a racoon caught in Milwaukee County in 1984 contained 1.4 mg/kg of DDT, 6.4 mg/kg of DDE, and 0.2 mg/kg of DDD (pesticides). However, there is no evidence that existing levels of contamination negatively impact on the abundance or diversity of wildlife species.

Loss of Wildlife Habitat

There is very little wildlife habitat in the Area of Concern. This is an urban area supporting a variety of commercial, industrial and residential uses. Thus, physical development has diminished wildlife habitat, rather than water quality problems or contamination.

RAP Preparation and Citizen Participation

The Wisconsin Department of Natural Resources (DNR) is responsible for developing the Milwaukee Estuary RAP, however advisory committees serve an instrumental role in preparing and implementing the plan. The Citizens Advisory Committee (CAC), the Technical Advisory Committee (TAC) and the Citizens Education and Participation Subcommittee (of the CAC) have contributed to RAP development and will assist in implementation of all recommendations. A tentative schedule for plan development is presented in Table 11.

The RAP's Technical Advisory Committee (TAC) brings together technical experts familiar with the Area of Concern to provide information needed to prepare and review the plan. Specifically, the TAC provides the following:

- problem identification, including impaired uses and their causes and sources (pollution, poor physical habitat, inappropriate riparian land use activities, eutrophication);

- technical problem analysis and assistance in defining preliminary goals and objectives for managing and restoring the resource;
- identification and evaluation of alternative approaches to managing and restoring the resource; and
- recommendations to implement preferred actions and restore the resource.

The TAC has identified problems and is developing management alternatives to control toxic substances, nutrient loadings, and ex-

TABLE 11.   The Schedule for Milwaukee's RAP Development Process

Planning Step	Anticipated Date for Completion
Technical Advisory Committee (TAC) established	May 1989
Citizen Advisory Committee (CAC) established	July 1989
CAC adopts "Desired Future State"	February 1990
CAC adopts preliminary goals and objectives; problem identification	May 1990
Public Meeting/Workshop on problem identification; preliminary goals and objectives	June 1990
Submittal of problem identification chapters of RAP to IJC (Stage I)	July 1991
Management Alternatives	
TAC Toxic Substances Monitoring Strategy	April 1992
TAC and CAC approve pollution abatement/management alternatives	April 1992
Public Hearing/Workshop on pollution abatement and management alternatives	July 1992
Recommended Plan Completed	
CAC approves final draft	August 1992
Public Hearing on final draft	September 1992
RAP submitted to IJC (Stage II)	October 1992

cessive eutrophication, as well as to improve biota, fish and wildlife populations, and habitat. TAC work groups have analyzed the pollution causes and sources of impaired uses for point sources and combined sewer overflows (CSOs), nonpoint sources, contaminated sediments, contaminated biota and human health, atmospheric deposition, leaching from landfills, surface water spills, and contaminated groundwater. A work group to evaluate dredging alternatives also has been formed. The reports of the work groups are used to draft various chapters of the RAP. The TAC is developing a Toxic Substance Monitoring and Management Strategy to interpret existing data, and propose a monitoring plan to identify and collect data required to make cleanup recommendations and evaluate progress in restoring beneficial uses.

In order to successfully prepare and implement the RAP, a cooperative effort among numerous and diverse interests is needed. Spearheading this vital coalition building effort for the Milwaukee Estuary RAP is the Citizen Advisory Committee (CAC), which includes local officials (city council alderpersons and County Board supervisors), local government agencies (staff from the city's Department of City Development and the Port of Milwaukee, County Department of Parks and Recreation and staff to the County board), sports fishing associations, commercial fishing associations, environmental interest groups (Sierra Club, Lake Michigan Federation), community action interest groups, civic groups and neighborhood associations (League of Women Voters, Wisconsin Action Coalition, Interfaith Conference of Greater Milwaukee, Keep Greater Milwaukee Beautiful, Milwaukee Urban League, East Side Housing Action Coalition, Historic Third Ward Association, Metropolitan Milwaukee Civic Alliance, Allied Council of Senior Citizens), the Milwaukee Metropolitan Sewerage District, the Public Policy Forum, the University of Wisconsin-Milwaukee, organized labor (Milwaukee County Labor Council and OSHA/Environmental Network), business groups (Greater Milwaukee Committee and Milwaukee Metropolitan Association of Commerce, Wisconsin Electric Power Company, Edelweiss Excursion Vessel, Miller Brewery), and the Southeastern Wisconsin Regional Planning Commission.

The CAC advises Wisconsin DNR on environmental restoration, and recreational and economic revitalization of the Milwaukee Estuary. Specifically, the CAC's charge is to do the following:

- represent the interests of key organizations and constituencies in the development of the Milwaukee Estuary RAP;
- review RAP chapters and reports from the TAC; and
- initiate public education programs that:
  a. familiarize citizens with impaired uses and the causes of poor water quality and habitat losses;
  b. demonstrate the loss of economic opportunities and low environmental value of the Milwaukee area because of pollution problems;
  c. promote citizen responsibility for restoring the Area of Concern and the entire river basin; and
  d. generate acceptance of recommended cleanup activities and provide motivation to implement these remedial measures;
- encourage and assist public participation in the RAP process, including the development of a vision for the Area of Concern, RAP goals, objectives, remedial measures, and implementation measures; and
- develop a strategy to implement the RAP's recommendations and unite the diverse interests necessary to successfully implement the RAP.

One of the CAC's first achievements was the development of the "Desired Future State," a vision statement that the committee has for the future of the estuary as a result of the successful implementation of the RAP's recommendations. The "Desired Future State" is defined as the following:

- waterways that, because of their purity, contribute to the economic vitality and quality of life in Milwaukee;
- waters, sediments, and biota that are free of persistent toxic or harmful substances resulting from industrial or other human activities—from the past, present or future;
- maximum public access and recreational opportunities along the rivers and nearshore areas of Lake Michigan for boating, swimming, fishing, hiking, bicycling, nature study and other leisure activities; and
- an estuary whose cleanliness and continued multiple uses have broad community and governmental support as a top priority of the local political agenda.

Public awareness of the problems in the Area of Concern, the RAP process and the need for remedial actions is critical to generate the "political will" to implement the RAP's recommendations. The "think-tank" for developing initiatives to build community-wide interest and support has been the RAP's Citizen Education and Participation Subcommittee (CEPS), a CAC subcommittee that assists in developing and implementing all RAP citizen education and participation efforts.

Throughout RAP development, public awareness is and will continue to be generated through several activities: public meetings will be held throughout the RAP planning process to obtain citizen input and review of the RAP goals, problem identification and recommendations for cleanup of contaminants; media coverage is being generated through periodic press briefings and other media events; similarly, regular briefings are being held for local public officials and business leaders. The CAC has toured the Area of Concern with local officials to review pollution problems and newsletters, brochures and fact sheets also have been developed to foster public interest. On April 30, 1990, the CAC sponsored a luncheon attended by 200 local business and civic leaders where Milwaukee's Mayor John Norquist and County Executive David Schultz issued a formal proclamation in support of the RAP's "Desired Future State" and restoration efforts.

In addition to these efforts, the RAP process is building on public participation efforts of Wisconsin DNR's other water quality programs in the basin. A survey of residents in the lower Milwaukee River basin was conducted in July and August 1989 to gauge public perception of water quality and support for cleanup programs. The survey was mailed to 5,500 residents and received an average response rate of 55 percent. Sixty percent of respondents rated local water quality as poor, while 55 percent of all Milwaukee residents believe that industry is the most important cause of water quality problems; less than 15 percent perceive nonpoint sources as significant causes of pollution. Only 9 percent use the Milwaukee River for boating and only 5 percent use it for fishing. About 35 percent of survey respondents (and the city, county, and Milwaukee Metropolitan Sewage District) say they are willing to recycle wastes and reduce their use of fertilizers and pesticides (the priority watershed nonpoint source programs and the integrated resource management plans). The most significant finding of the survey is that more than 50 percent of survey respondents are willing to pay more in taxes or sewer fees for pollution abatement efforts. This information must be used by the RAP to generate funding support for implementation.

Specific Remedial Actions Planned and in Progress

The Milwaukee Estuary RAP builds on the progress and accomplishments of numerous pollution abatement and planning efforts in the Milwaukee River basin. Some of the more noteworthy cleanup efforts are discussed below.

Milwaukee Metropolitan Sewerage District

The Milwaukee Metropolitan Sewerage District's (MMSD) Water Pollution Abatement Program is a long-range program designed to significantly reduce point source pollution discharges and greatly improve wastewater treatment (MMSD 1990). This program entails a massive renovation and expansion of the sewerage system and treatment plant capabilities. At a project cost of more than $2.2 billion, the MMSD project is the largest and most expensive wastewater treatment upgrade and rehabilitation project in Wisconsin's history. The project includes renovation of the Jones Island and South Shore Wastewater Treatment Plants (WWTPs), rehabilitation of sewer lines, alleviation of overflows from the separate sewer system, and installation of relief sewers (also called the "deep tunnel" project) to significantly reduce combined sewer overflows and the bypass of wastewater to the surface water. The relief sewer construction project includes approximately 27 km (17 miles) of deep tunnels, up to 9.8 m (32.3 feet) in diameter, to store untreated wastewater during wet weather conditions.

The Jones Island Wastewater Treatment Facility has maintained continuous permit compliance for over 10 years, and the bypassing of sewage has been significantly reduced through recent construction improvements. Upon completion of the project, overflows from the separate sewer system will be eliminated. Approximately 109 combined sewer overflows impact the Menomonee, Kinnickinnic, and Milwaukee Rivers, as well as the turning basin and the outer harbor. Combined sewer overflow discharges occur when Milwaukee receives more than 0.25 cm (0.1 inch) of rain or snow and can occur between 50 to 70 times per year. By 1993, discharge from combined sewer overflows will be reduced to one or two occurrences per year. The surface waters impacted by these pollutant sources include the Menomonee, Kinnickinnic, and Milwaukee Rivers, and the inner and outer harbors. Significant reductions in point source pollution and corresponding improvements in water quality will occur when MMSD completes its entire water pollution abatement program in 1996.

Southeastern Wisconsin Regional Planning Commission

In 1979, the Southeastern Wisconsin Regional Planning Commission (SEWRPC) prepared the Areawide Water Quality Management Plan for southeastern Wisconsin. Required by Section 208 of the Clean Water Act, this plan identified water resources management recommendations for use by state and local governments to guide the use of waterways and the development along the streams and lakes of southeastern Wisconsin. Wisconsin DNR's southeast district headquarters has updated the portion of the plan that addresses the Milwaukee River basin through Integrated Resource Management Plans (discussed in the following section).

In 1987, SEWRPC prepared *A Water Resources Management Plan for the Milwaukee Harbor Estuary* (the Milwaukee Harbor Estuary Study) in cooperation with MMSD, the U.S. Geological Survey and Wisconsin DNR (SEWRPC 1987) to review the adequacy of MMSD's ongoing water pollution abatement program and to develop additional recommendations to solve water quality problems within the Milwaukee Harbor. The plan also considered how to reduce damage caused by flooding, storm and wave action; provide for navigation by deep draft vessels through implementation of an environmentally-sound maintenance dredging program; prevent shoreline deterioration in the harbor; and maximize the harbor as a prime urban recreation area. The study presents recommendations to resolve water quality problems in the Area of Concern for conventional pollutants and concludes that further study is needed to address toxic substances issues. It provides an extensive data base that has greatly benefited the RAP planning process.

Integrated Resource Management Plans and Priority
Watershed Program

On May 8, 1984, Governor Anthony S. Earl signed Wisconsin Act
416 into law, which created the Milwaukee River Priority Watersheds Program and appropriated between $3.1 and $7 million to
landowners and local governments to clean up nonpoint source pollution. Wisconsin DNR expanded the priority watershed program
to include a more comprehensive planning effort that addresses all
categories of water quality problems. This includes two major components: the Nonpoint Source Priority Watershed Plans (PWPs) and
the Integrated Resource Management Plans (IRMPs).

The PWPs provide a voluntary approach to nonpoint source pollution control, including up to 70 percent in cost-sharing assistance
to eligible landowners and local governments willing to install best
management practices to reduce water quality impacts (Wisconsin
DNR 1990). The IRMPs, which update the Areawide Water Quality
Management Plan for the Milwaukee River Basin, integrate the responsibilities of Wisconsin's nonpoint source and land management
programs with the work of other environmental quality programs
(solid waste, wastewater, water regulation, water supply, environmental analysis and review) and resource management programs
(fisheries, forestry, parks and recreation, water resources, wildlife
and endangered resources). IRMP objectives and recommendations
serve as guidelines for state and local governments to guide future
development and use of natural resources in the Milwaukee River
basin (Wisconsin DNR 1990).

The five priority watersheds designated by the legislature for
cleanup are the Milwaukee River East-West Branch, Milwaukee
River North Branch, Cedar Creek, Menomonee River and the Milwaukee River South. These five watersheds ultimately flow to the
Area of Concern and collectively encompass about 2,150 km² (830
square miles), including approximately 656 km (410 miles) of peren-
nial streams and 84 lakes and ponds 2 hectares (5 acres) or larger.
The Nonpoint Source Priority Watershed and Integrated Resource
Management Plans for the East-West Branch and North Branch Watersheds were completed in December 1988 and June 1989, respectively, while the Nonpoint Source PWPs and Integrated Resource
Management Plans for the Milwaukee River South and the Meno-

monee River Watersheds were completed in June 1990. The IRMP for Cedar Creek will be completed in April 1990 and the Nonpoint Source Priority Watershed Plan should be completed by December 1990. Implementation will begin for all plans in 1990.

Milwaukee River Revitalization Council

To assist Wisconsin DNR in achieving improved water quality in the Milwaukee River basin, the Wisconsin Legislature created the Milwaukee River Revitalization Council in 1987. The council advises the state on matters relating to the economic, recreational and environmental revitalization of the Milwaukee River basin.

Harbor Strategic Plans

The Milwaukee Harbor Strategic Plan (Page et al. 1989), developed by the University of Wisconsin-Milwaukee and funded by the Wisconsin Coastal Management Program, is a long-term strategic plan for the Milwaukee Harbor and connecting waterways. The plan examines the maximum environmental, economic, and recreational potential for the estuary, assuming environmental restoration of water resources and rehabilitation of adjacent lands.

The Port of Milwaukee's Strategic Plan for 1988–1993 provides guidance for future port activity based on operations, market changes, and current as well as future business options. Objectives to be achieved by 1993 include the port's development as a regional transportation and business center, infrastructure renewal, and financial self-sufficiency.

City of Milwaukee River Development Guidelines

The City of Milwaukee is implementing a river development program and developing a master plan for the Menomonee Valley. The river development project will help shape the Milwaukee River's role as a major tourist attraction through the strategic placement of dining areas on river rafts, waterway transportation and taxi stops, public docking facilities and increased public access, and a complete riverwalk system. The project should increase public interest and use of the estuary, thereby validating the RAP's efforts to further improve water quality. The Milwaukee Department of City Development also is considering land use development guidelines that will be consistent with the water quality and habitat restoration initiatives of the RAP. As part of its Menomonee Valley Master Plan, the City of Milwaukee is considering creation of an 11 hectare (27 acre) wetland contiguous to the Menomonee River and within the Area of Concern to provide valuable fish, aquatic life and wildlife riverine habitat as well as flood control and water quality benefits.

County Riverland and Lakefront Plans

Milwaukee County also is proceeding with important initiatives. The Milwaukee County Riverland Acquisition, Preservation and Development Study examines acquisition and development of riverine lands in Milwaukee County in accordance with the principles of environmental and resource protection, public investment protection, economic development and neighborhood stability. Similarly, Milwaukee County has developed its Lakefront Park System North Harbor–McKinley–Lake Park Sites Masterplan. This strategy plans for long-range capital investment in recreation programs, seeks to protect existing Milwaukee County investments in the lakefront park system, and promotes aesthetic and functional compatibility between projects.

Additional Monitoring/Research Needs

While information is available from several years of research, more information about toxic contaminants, their transport and interaction in the Milwaukee Area of Concern is necessary. A monitoring strategy is being developed as part of the RAP to determine where research is needed to learn more about the persistent contaminants in the system. Major components of the strategy include the following:

- A mass balance to determine the relative contributions of different pollution sources so that cleanup activities can be prioritized. Contaminated sediments and runoff are major sources of toxic substances, but little is known about the contribution of abandoned and active landfills. More information is needed to determine which contaminants exist in runoff from storm sewers and industrial sites.

- Further monitoring to identify contaminated sediment "hotspots" and quantify their impact on the food chain.
- A comprehensive survey of the existing fishery to document existing diversity and establish baseline measures for future improvements.
- A study to determine if waterfowl are picking up significant levels of contaminants from the Area of Concern (from the confined disposal facility or other sites within the estuary).
- Finally, information on the effects of contaminants on reproduction and survival of fish and wildlife.

Challenges to the Full Implementation of Remedial Actions

At least five major challenges face the Milwaukee Estuary RAP, each of which underscores the need for a comprehensive and well-coordinated effort involving the federal, state, provincial, and local governments.

Adequate Funding

Foremost is the need for adequate funding for research, monitoring, and pollution abatement efforts. Costs for cleanup efforts are not known, but will likely be significant. Current RAP planning efforts in Milwaukee involve the careful nurturing of a public-private partnership between the State of Wisconsin, Milwaukee County, the City of Milwaukee, the Milwaukee Metropolitan Sewerage District, and the private sector. While this partnership will continue through implementation, fiscal constraints and the political vulnerability of local and state budgets could impede progress. Property taxes are the principal source of income for Milwaukee County and the City of Milwaukee, and environmental cleanup programs are competing with programs addressing other social and economic needs for scarce local dollars. At the same time, household incomes are declining, largely due to the recession of Milwaukee's manufacturing economy.

Public Awareness

The efforts of the Citizens Advisory Committee and its Citizen Education and Participation Subcommittee must continue throughout and beyond the RAP implementation period. Citizen education and public official briefings must continue to ensure that policy priorities remain focused on cleanup of the rivers and harbor. Continued local involvement also is necessary to maintain sufficient funding for cleanup efforts.

Political Support and Implementation Leadership

Implementation of the Milwaukee Estuary RAP recommendations must become a top priority among public officials. Milwaukee's Mayor Norquist has signed a proclamation publicly stating his support for the RAP and its implementation, and his proclamation indicates that local officials are becoming involved and are listening to citizens who want a clean, safe environment. The Milwaukee Estuary RAP Citizen Advisory Committee includes the mayor's representative, the county executive's representative, city alderpersons, county supervisors, and state legislators. In addition to strategies for building community-wide awareness, the RAP effort will continue to integrate local leaders in the planning effort to build local government ownership for implementation of remedial actions.

Necessary Rules and Regulations to Guide
Implementation

Federal and state governments must provide the incentives and regulatory tools for successful implementation to become a reality. The Governments of Canada and the United States, through the International Joint Commission and by acts of government legislative bodies, must guide and mandate more direct action for implementation of zero discharge policies for persistent toxic substances. The federal and state governments also must provide standards for sediment-associated contamination, regulations for their remediation and disposal, and additional phosphorus controls.

The State of Wisconsin has moved through implementation of

rules to restrict the discharge of toxic substances. This permit program contains special provisions to protect the Great Lakes from toxic compounds that bioaccumulate. Through antidegradation rules, the state is preventing increases in the discharge of persistent toxic substances and restricting toxic air emissions that should reduce atmospheric deposition to the Great Lakes from Wisconsin's stacks.

However, more input from state and federal government is needed to assist in RAP implementation. One good example is the Great Lakes Critical Programs Act of 1990 which mandates the development of all United States RAPs by January 1, 1992, authorizes additional resources to ensure timely RAP development and mandates that each RAP is to be included in each state's Water Quality Management Plan.

Advanced Technologies

A significant challenge facing the Milwaukee RAP is the inadequacy of current technologies. The only proven (and most economically-feasible) technology to address contaminated sediment is to excavate the dredge spoils and deposit them elsewhere, such as in a confined disposal facility or a hazardous waste storage facility. This solution often poses additional environmental problems because containment of contaminants has not always proven successful.

Pilot efforts are underway in Wisconsin's Sheboygan River RAP to remediate PCB contaminated sediments. Three contaminated sediment hotspots are being dredged; these sediments are being placed into a confined treatment facility. Other areas of contamination are being armored to isolate them from contact with the river environment. These efforts are experimental, but may provide effective options to remediate contaminated sediments in the Milwaukee estuary.

More research is needed to develop advanced and economically feasible technologies that address such concerns as in place pollutants, urban runoff, and atmospheric deposition. Waste minimization and source reduction also should be used to prevent further contamination.

Conclusion

Stage 1 of the Milwaukee Estuary RAP is now complete and Stage 2 has begun. Implementing the RAP and confirming restoration of beneficial uses may take several decades. However, the process has drawn support from many interests in the Milwaukee Area of Concern. Implementation poses challenges that demand cooperation from all government sectors, as well as the support and backing from citizens, local industries and businesses, and other interest groups.

REFERENCES

Milwaukee Metropolitan Sewerage District (MMSD). 1990. *Water Pollution Abatement Program*. Milwaukee, Wisconsin.
Page, G.W., J.V. Klump, H.M. Mayer, and S.B. White. 1989. *The Milwaukee Harbor Strategic Plan*. U. W. Wisconsin-Milwaukee. Center for Great Lakes Studies. Wisconsin Coastal Management Program, Milwaukee, Wisconsin.
Southeastern Wisconsin Regional Planning Commission (SEWRPC). 1987. *A Water Resources Management Plan for the Milwaukee Harbor Estuary*. Planning Report No. 37, Volumes 1 and 2, Waukesha, Wisconsin.
Wisconsin Department of Natural Resources. 1990. *Nonpoint Pollution Abatement and Integrated Resource Management Plans for the Milwaukee River South and Menomonee River Priority Watershed*. Milwaukee, Wisconsin.

Chapter 8

An Overview of the Modeling and Public Consultation Processes Used to Develop the Bay of Quinte Remedial Action Plan

Frederick Stride, Murray German, Donald Hurley, Scott Millard, Kenneth Minns, Kenneth Nicholls, Glenn Owen, Donald Poulton, and Nellie de Geus

"The Quinte RAP is the means to cleanup and, once the cleanup has been completed, sustain Bay of Quinte water quality. The RAP is a holistic approach which has involved all economic sectors, levels of government and the public. Realistic and attainable goals have been established. As well, the Quinte RAP has broadened the perspective of the players and provided the first steps to cooperative environmental protection. The Quinte RAP truly reflects the view, goals and environmental aspirations of the Quinte area residents."

Glen Hudgin
Chairperson, Bay of Quinte
RAP Public Advisory Committee

Introduction

The Bay of Quinte was designated a "problem area" by the International Joint Commission in 1975 because of excessive nutrient enrichment, nuisance algae growth, low concentrations of dissolved oxygen in bottom water, and localized bacterial contamination. These problems were apparent well before that time (Johnson and Owen, 1971; McCombie 1967) and abatement activities were taken: upgrading sewage treatment plants, limiting the phosphorus content of laundry detergents, and reducing phosphorus concentrations in sewage treatment plant effluents to 1.0 mg L^{-1} (annual average) in 1975 and 0.5 mg L^{-1} (May to October average) in 1978. As a result, water quality conditions generally improved (Minns et al. 1986a and b).

In 1985, a review of existing data showed that nine of the 14 beneficial uses identified in Annex 2 of the Great Lakes Water Quality Agreement were still impaired in the Bay of Quinte. Primary problems include the following: nuisance algal growth, localized bacterial contamination, toxic contamination, restrictions on fish consumption, and loss of aquatic habitat. To address these concerns the remedial action plan (RAP) process for the Bay of Quinte was initiated in 1986. A RAP team was formed with representation from Environment Canada, Environment Ontario, Fisheries and Oceans Canada, the Ontario Ministry of Agriculture and Food, and the Ontario Ministry of Natural Resources. Each environmental problem was identified with greater precision as were the sources of the problems and potential remedial actions. The success to date of the Bay of Quinte RAP is attributable to three factors:

1. Project Quinte, a long-term, multiagency evaluation of ecosystem responses to phosphorus loadings in the Bay of Quinte before and after phosphorus inputs were reduced in 1977–78;
2. The development and use of phosphorus and toxic contaminant models as tools to evaluate and select potential remedial actions; and
3. An extensive public education/consultation program.

These factors, their application to ecosystem restoration in the Bay of Quinte, and the institutional framework for implementation of the Bay of Quinte RAP are discussed in this chapter.

Background

Physical and Biological Description

The Bay of Quinte is part of the northeastern shore of Lake Ontario separated from the main lake by Prince Edward County (Figure 13). The bay is Z-shaped and includes three smaller areas commonly referred to as the upper, middle, and lower bays. The Bay of Quinte is approximately 64 km (40 miles) long with an area of 254 km² (98 square miles). Water depths increase from west to east, ranging from as shallow as 1 m (3.3 feet) in the extreme western end of the upper bay to a maximum depth of 66 m (216 feet) in the lower bay. The

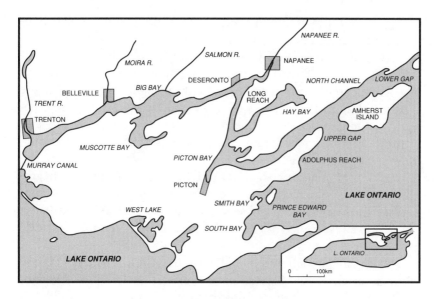

Fig. 13. The Bay of Quinte

maximum width of the bay is 3.5 km (2.2 miles) (Sly 1986) and the drainage area is 17,315 km² (6,685 square miles).

Lake Ontario water enters the lower and middle bays at depth, exchanging and mixing the waters between Lake Ontario and bay water. Some Lake Ontario exchange also occurs at the Murray Canal in the upper bay. Tributary flows strongly influence the upper bay's dilution capacity, flushing rate, and water quality. Generally, the entire bay's water is flushed through the bay two to three times per year. Flushing rates vary between each section and decrease significantly between May and September, then increase five to 10 times during spring snow melt (Minns et al. 1986b).

Major tributaries include the Trent, Moira, Salmon and Napanee Rivers. The Trent River watershed is the largest and accounts for 67 percent of the total drainage area. It extends north to Algonquin Park and east to within 25 km (15.6 miles) of Lake Simcoe. A series of dams, locks and canals—the Trent Severn Waterway—provide a navigable route on the Trent River. A smaller canal, the Murray Canal, links the western end of the upper bay to Lake Ontario (Sly 1986).

Significant alterations in species composition and abundance of fish have occurred in the bay in recent decades (Hurley and Christie, 1977). Some changes during the 1950s and 1960s (reductions in walleye, lake whitefish, and several centrarchids) have been linked to the white perch invasion and increasing eutrophication. During the 1960s and up to 1978, stocks of alewife and white perch were especially abundant, while northern pike, bowfin, longnose gar, and walleye were near record lows. Eutrophication impaired the production of piscivorous fish in the Bay of Quinte while enhancing the production of smaller-bodied planktivores and benthivores. Beginning in 1977 (coincident with reduction in phosphorus inputs from sewage treatment plants and severe winters in both 1977 and 1978), there was a resurgence of the walleye stock and a collapse of white perch.

The fishable population of walleye today is estimated at 700,000 (Bowlby et al. 1990). Alewife have declined in the upper bay, but decreased in the middle and lower bays only in the late 1980s, in response to walleye predation and intraspecific competition (Ridgeway et al. 1990). The numbers of alewife are now the lowest seen since the early 1960s.

Toxic contaminants are found in seven fish species such that restricted consumption is recommended for larger sizes. Mercury, polychlorinated biphenyls (PCBs), and mirex are the most frequently found contaminants, while dioxins and furans have been recorded in low concentrations in larger walleye.

More detailed information on the conditions of the Bay of Quinte can be found in Minns et al. (1986a) and Sly (1986).

Land Use and Development

The first European settlers arrived in the late 1800s. By 1830 land near the bay had been cleared for commercial agriculture, and mining and quarrying operations also began. Timber production peaked in the 1860s (Sly 1986, Ecologistics unpubl.). Agricultural activities (corn production and dairy herding) today account for 37 percent of land use activities in the Bay of Quinte watershed, while 59 percent of the basin is forested (Michalski 1987).

Five population centers are in the watershed: Belleville (population 35,000), Trenton (15,000), Napanee (4,800), Picton (4,200) and

Deseronto (1,700). Nine townships also border the bay with a combined population of about 40,000.

Water Use

The Bay of Quinte provides water for Belleville, Picton, Deseronto and, in the near future, Sidney Township, serving 68,000 people in the bay area (Ecologistics, unpubl.). Treated municipal and industrial wastes are discharged into the Bay of Quinte from all five population centers and the Canadian Forces Base in Trenton. Extended aeration processes are used at Deseronto, while activated sludge techniques are utilized elsewhere in other water treatment plants. The capacity of all sewage treatment plants is 93,400 m^3 day^{-1} with a total daily average flow of 78,000 m^3 day^{-1}.

No industries discharge wastes directly to the bay. Two paper mills (Domtar Packaging and Trent Valley Paperboard) and a wood-preserving operation (Domtar Wood Preserving) discharge treated wastes into the Trent River and a third paper mill (Strathcona) discharges treated waste to the Napanee River. A distillery (Corby) also releases treated waste to the Moira River. While Bakelite Thermosets in Belleville has ceased operations, contaminants enter the bay from this site in surface and groundwater runoff.

Eastern Lake Ontario, including the Bay of Quinte, supports a valuable commercial fishery. Commercial catch totalled 540 metric tonnes in this region in 1988 at an approximate value of $897,000 (Ontario Ministry of Natural Resources 1989). However, the industry and total tonnage caught have declined in recent years. This decline may be attributable to lower market prices, increasing costs and increased management of the sport fishery.

The Bay of Quinte sports fishery is thriving because of the availability of walleye—96 percent of the anglers in 1988 sought this species. In that year alone 114,300 walleye were caught in open water, although the harvest has varied between 90,000 and 130,000 walleye since 1982 (Ontario Ministry of Natural Resources 1989).

Other summer and winter recreational activities are provided in the bay. Six swimming areas are located on the bay as well as a number of smaller, local beaches. However, these swimming areas and beaches are frequently closed due to bacteriological contamination (Poulton 1989; Ecologistics unpubl.).

Project Quinte

The primary goal of Project Quinte was to examine ecosystem responses to reduced phosphorus loadings from Bay of Quinte municipal sewage treatment plants. Seasonal cycles of nutrient loading, hydrology, and the biotic communities were examined to describe productivity at each trophic level, to compare long-term changes and their possible causative factors. The geographic gradients in trophic status from the upper to the lower bay were examined and showed a range in the values of conditions such as phosphorus supply, water temperature, and mixing depth, all of which affect environmental conditions and ecosystem status. As a result, the role of these controlling factors could be determined and the merits of phosphorus management in different segments of the bay assessed.

Project Quinte has been a multiagency undertaking. Principal agencies involved include Environment Ontario (Aquatic Ecosystems Section–Rexdale and Technical Support Section–Kingston), Ontario Ministry of Natural Resources (Ontario MNR Glenora Fisheries Station–Picton), Fisheries and Oceans Canada (DFO Burlington) and Queens University (Kingston). Each agency assumed responsibility for establishing and maintaining various parts of the data base. For example, chemical and physical limnological data, and biomass and production of phytoplankton and zooplankton data were collected by Environment Ontario (Rexdale) and DFO. Measurement of nutrient loadings and phosphorus budgets was an activity shared by Environment Ontario (Kingston) and DFO, while higher trophic levels were sampled by DFO (benthos), Queens University (macrophytes), and Ontario MNR (fish).

As a result of the many studies conducted under Project Quinte, significant insight on how the bay functions and responds to change was achieved and several facts important to the management of the ecosystem were ascertained. In response to a 50 percent decline in point source phosphorus loadings in 1978, average May-to-October phosphorus concentration declined 35 percent in the upper bay, 20 percent in the middle bay, and remained unchanged in the lower bay. In spite of substantial reductions in the upper bay, algal biomass fluctuated widely and 1987 and 1988 concentrations approached levels typical of prephosphorus controls (Project Quinte unpubl.). No significant changes in algal community composition have been observed, and diatoms and blue green algae continue to dominate (Nicholls et al. 1986). At the same time, primary production rates and chlorophyll *a* have been reduced by 50 percent in the post-1978 period (Millard and Johnson, 1986).

The zooplankton community is dominated by species typical of eutrophic conditions, such as small cladocerans. Species composition remains unchanged despite phosphorus management and the ecosystem changes noted above. Macrophytes were found in only 14 percent of the bay in the 1970s. No significant increase in distribution or abundance occurred in the phosphorus control period (Crowder and Bristow, 1986).

Within the benthic community significant changes have occurred in the post-phosphorus control period. In response to increased predation and declining primary production, the numbers of certain benthic organisms (oligochaetes, chironomids, and sphaeriids) have decreased (Johnson and McNeil, 1986). At the same time, the populations of more pollution tolerant species have increased.

Significant changes in fish communities have occurred since 1978. Coinciding with phosphorus reduction, white perch declined, alewife populations fluctuated widely, and walleye increased (Hurley et al. 1986). Community diversity remains low, and an apparent lack of stability in the fish community exists.

RAP Development

The historic data base generated through Project Quinte was used to set short-term priorities and define realistic restoration goals for the RAP. As a result, immediate action was focused on nutrient enrichment, the area of greatest scientific understanding. The linkages between forcing factors (e.g. climate, nutrient loads, etc.) and ecosystem responses also had been identified, which allowed the RAP team to discuss conventional pollution abatement activities and "ecosystem approaches." A data base and monitoring program is required to complete the RAP process, and the Project Quinte experience provided the basis for the monitoring effort and offered valuable insight and direction.

Past studies of nutrient types, amounts and dynamics in the bay (Johnson and Owen, 1971) showed that numerical assessment modeling of phosphorus input-outputs could be done. A similar assessment of the Bay of Quinte, reflecting the ecosystem approach, was desired. The RAP team's objectives were to integrate present information and knowledge about the ecosystem and to provide an objective evaluation of all potential remedial options.

The modeling process (construction of a computer simulation model of the Bay of Quinte ecosystem) was accomplished by drawing on the expertise of Quinte researchers. An adaptive environmental assessment and modeling (AEAM) methodology (Holling 1978) was used. Various subcomponents of the model were developed, uncertainties identified, and policy options explored. Thirty-nine people, including many experts, long-time Quinte researchers and members of the Bay of Quinte RAP Public Advisory Committee (PAC), participated.

A detailed numerical model of phosphorus dynamics in the Bay of Quinte, particularly the upper bay, was constructed. In addition, a preliminary PCB model of the bay and a conceptual model of biotic interactions also was developed. Twelve phosphorus control scenarios were considered in the phosphorus ecosystem model, including phosphorus loading reductions from sewage treatment plants, industrial sources, tributaries and water treatment plants, and a scheme to divert water from Lake Ontario into the upper bay ("flushing option"). An option for no further actions beyond those already completed also was evaluated (Table 12).

The flushing option, while the least responsible from an ecosystem perspective, was the most effective single action. However, a combination of remedial actions applied to the maximum extent possible in each sector was predicted to reduce substantially the phosphorus levels in the upper bay. Lastly, the model predictions demonstrated the central role of sediment recycling; sediment phosphorus has decreased and, as a result, phosphorus concentrations in the water column should continue to improve for the next 10 to 20 years without further external loading reductions (see Figure 14—Option #1 "No Further Remedial Actions").

The structured, conceptual model of biotic interactions contributing to ecosystem imbalance produced several hypotheses. While numeric representation was not constructed, the "cause-effect" web was documented and evidence for and against the suggested linkages was compiled. Results of this exercise emphasized the

TABLE 12.   Results of Phosphorus Control Measures for the Years 1995 and 2005 (base year 1990 in both cases)

	Results 1995		
Option	P Concentration (μg/L)	Algal Biomass (mm^3/L)	Potential Macrophyte Area (km^2)
1	31.7–42.7	5.9–8.0	39.6–37.0
2	30.7–41.2	5.7–7.7	39.8–37.3
3	29.8–39.5	5.5–7.4	40.1–37.7
4	31.1–41.5	5.8–7.7	39.7–37.2
5	30.8–41.1	5.7–7.7	39.8–37.3
6	30.4–40.7	5.6–7.6	39.9–37.4
7	29.7–40.7	5.5–7.6	40.0–37.4
8	28.5–39.4	5.3–7.4	40.0–37.7
9	29.4–37.6	5.5–7.0	40.1–38.1
10	28.0–34.7	5.2–6.5	40.5–38.8
11	31.3–42.1	5.8–7.9	39.7–37.1
12	23.9–29.3	4.4–5.4	41.6–40.2

	Results 2005		
Option	P Concentration (μg/L)	Algal Biomass (mm^3/L)	Potential Macrophyte Area (km^2)
1	29.8–38.9	5.5–7.3	40.1–37.8
2	28.3–37.1	5.3–6.9	40.7–38.7
3	27.2–35.3	5.1–6.6	40.7–38.7
4	28.6–37.5	5.3–7.0	40.3–38.2
5	28.3–37.1	5.3–6.9	40.4–38.3
6	28.0–36.7	5.2–6.8	40.5–38.7
7	27.3–36.8	5.1–6.8	40.1–38.3
8	26.1–35.4	4.8–6.6	41.0–38.7
9	27.3–34.2	5.0–6.4	40.7–38.9
10	26.0–31.6	4.8–5.9	41.0–39.6
11	28.9–38.3	5.4–7.1	40.3–38.0
12	21.8–25.8	4.0–4.8	42.2–41.1

Note: The 12 options are as follows:

Option 1: no further remedial actions beyond those actions already taken
Option 2: STP effluent controls — summer effluent P concentration 0.3 mg/L; yearly 0.7 mg/L
Option 3: STP effluent controls — summer effluent P concentration 0.1 mg/L; yearly 0.5 mg/L
Option 4: industrial P load — 70% of 1988 levels
Option 5: industrial P load — 40% of 1988 levels
Option 6: eliminate stormwater discharges
Option 7: tributary P loads reduced by 10%
Option 8: tributary P loads reduced by 16%
Option 9: flushing flow of 20 m^3/sec into the Upper Bay
Option 10: flushing flow of 35 m^3/sec into the Upper Bay
Option 11: eliminate direct discharges of Water Treatment Plant Sludge
Option 12: combination of Options 2, 4, 6, 7, 10 and 11

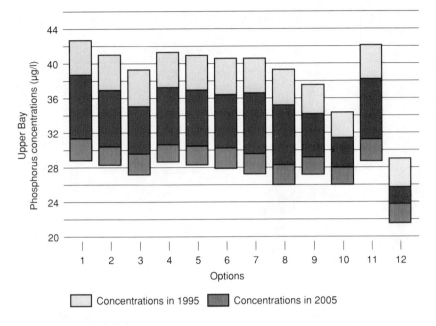

Fig. 14. Results of the 12 control option scenarios. Shown are the ranges of phosphorus concentrations in the Upper Bay under high and low flow regimes for the years 1995 and 2005. Cross-hatching shows area of overlap. (Source: ESSA 1988.)

need for more extensive numerical simulations of the Bay of Quinte ecosystem that integrate nutrients, contaminants, physical habitat, and biotic interactions.

The RAP team and the PAC were encouraged by the products of the Phosphorus-Ecosystem Model. The modeling exercise brought together many disparate pieces of the Bay of Quinte puzzle and fostered a greater sharing of knowledge and ideas among those contributing to RAP development. The model predictions were instrumental to RAP team/PAC preparation of the *Time to Decide: A Discussion Paper* report (Bay of Quinte RAP Coordinating Committee and Bay of Quinte RAP Public Advisory Committee 1989) and a comprehensive evaluation of remedial options for public consideration.

A fate-and-transport model of toxic contaminants in the Bay of Quinte, similar to that envisioned in the phosphorus-ecosystem modeling exercise, has been developed. As a first step, the model has been calibrated for arsenic and pentachlorophenol, the choice of which was designed to highlight the wide range of properties influencing the fate and transport of contaminants. For example, arsenic has little affinity for biota or sediments, while pentachlorophenol degrades rapidly in sunlight. These chemicals also were selected based on past use in the area: arsenic was used in past mining activities in the Moira River watershed (Mudroch and Capobianco, 1980) and pentachlorophenol is used as a wood preservative at Domtar Wood Preserving in Trenton.

Further development of the model will include additional contaminants, including heavy metals and organics, seasonal dynamics, and linkages to nutrient modeling. Model validation is limited by data availability, including inadequate detection limits for many contaminants. However, qualitatively assessing the relative importance of local and nonlocal sources suggests that remedial options can be evaluated despite the uncertainties. The modeling work continues and the impact of various abatement scenarios proposed as part of the Bay of Quinte RAP will be explored.

The RAP team's success with model construction provided the basis for further model work toward the eventual goal of an "Integrated Bay of Quinte Ecosystem Management Model." By developing models of the various features and concerns of the bay, in fact, the RAP team has built submodels for the integrated system. Another project, predicting potential improvements in macrophyte abundance to distribution and potential responses in fish assemblage, was initiated because earlier models indicated a lack of spatial resolution. A consulting firm is using a combination of modeling and geographic information systems (GIS) to examine spatial aspects of ecosystem response to remedial action.

This project will provide improved estimates of changes in the macrophyte area resulting from increased water clarity and, as a next step, predictions of potential changes in adult summer habitat for northern pike. The project, in turn, lays the groundwork for projecting changes in potential habitat for many important fish species and, potentially, as a tool for shoreline management practices. The RAP team also is building a model of the major biotic interactions in the Bay of Quinte ecosystem that involves algae-macrophyte interactions, interactions among piscivores, and between pis-

civores and habitat features, top-down/bottom-up processes determining the abundance of algae, and other interactions. In conjunction with earlier models, this model will provide the basis to assess
phenomena such as the zebra mussel invasion.

For application as a management tool, the Bay of Quinte RAP
team must link the various submodels together to produce interactive simulation models of nutrient dynamics, food-chain interactions, and contaminant dynamics for the Bay of Quinte. In this
way, output can be shared between models and/or a forcing variable
in one case can be substituted as a state variable in the next. For
example, the nutrient simulation model could be used in the food-
chain model as a forcing function of algal growth. In the long term,
the model will provide an ongoing basis to introduce monitoring
data, identify information and research needs, provide predictive
capabilities, and integrate the selection and assessment of remedial
actions.

Public Education and Consultation

The Bay of Quinte RAP Public Education and Consultation program
is essentially three separate but integrated and ongoing components: (1) information distribution, (2) structural organization and
(3) discussion and consultation. This structure provides for an ongoing exchange of information and allows for the program to be developed and/or enhanced. The public has a perceived role to play, this
role is meaningful and active, and has helped to build consensus for
the Bay of Quinte RAP and for local ownership of the plan.

The program began formally in the fall of 1986, when a general
public meeting was convened and the RAP process was outlined.
The public also voiced its concerns and perceptions about the local
water quality at the meeting. These views helped in establishing
the information distribution component of the program and, as a
result, communication, education and consultation actions as well
(Table 13). A Public Advisory Committee (PAC) was formed to provide participation and leadership in many Bay of Quinte RAP undertakings, to coordinate RAP team and PAC activities, to co-produce
the *Time to Decide* report, and to provide the ongoing commitment
to education and information dissemination on the part of the RAP
team and PAC.

TABLE 13.   The Bay of Quinte RAP Public Education/Consultation Actions (in chronological order)

Creation of a stakeholder mailing list

Creation and distribution of a Bay of Quinte RAP newsletter

Stakeholder cleanup and/or restoration goals established

Identification of the "impaired beneficial uses" in the Area of Concern by stakeholders

Public distribution of the Bay of Quinte RAP *Progress Report (February 1987)* — a summary of the identified water quality problems in the Area of Concern and a synopsis of the Project Quinte data base

Release of all Bay of Quinte RAP reports and technical studies (10) to the public via local public libraries and local government agencies

Public education sessions held following the release of each Bay of Quinte RAP technical report

Information and display booth located at local, water-related events (e.g. waterfront festivals and celebrations, fall fairs, etc.)

Presentations to local interest groups and schools by either RAP team members or PAC representatives

PAC and stakeholder participation (five members) in the Phosphorus–Ecosystem Modeling exercise (Summer 1987)

Formation of the Bay of Quinte RAP PAC in the spring 1988 to develop the Bay of Quinte RAP, to review RAP team undertakings, and to disseminate information to the stakeholders and the public-at-large

PAC review of all draft reports and in preparation Bay of Quinte RAP technical information

PAC representation on the Bay of Quinte RAP team (spring 1988)

Preparation of a PAC initiated video, *Time to Decide* (summer 1989)

Public distribution of the Bay of Quinte RAP team and PAC report *Time to Decide* summarizing the technical evaluation and remedial options to control phosphorus loading evaluated to date (fall 1989)

Four public open houses in fall 1989 to discuss the *Time to Decide* information

In December 1989, a PAC sponsored public meeting to receive stakeholder input to the "preferred remedial actions" described in the *Time to Decide* report

PAC review of Quinte RAP Stage 1 Report (January 1990)

(continued on next page)

(Table 13—*continued*)

PAC and stakeholder recommendations presented to the RAP team (March 1990)

PAC submission to RAP Steering Committee concerning implementation (April 1990).

PAC releases *1990 PAC Report* (April 1990) summarizing public response and input to the *Time to Decide* report

PAC forms Implementation Committee to oversee and participate in the implementation of the Bay of Quinte RAP recommendations for action

Stage 1 Report presented to the IJC's Restoration Subcommittee at the Glenora Fisheries Station, Bay of Quinte (July 1990)

The magnitude of the public education and consultation program is considerable: stakeholders total about 1,000 persons; 10 public meetings and four open houses have been convened; and presentations made to interest groups number over 200. Moreover, the PAC has 21 continuing volunteer members with alternates at each position and with representation from industry, environmental organizations, municipalities, the Tyendinaga Nation, and other interest groups.

In many ways the PAC and RAP team perform similar duties and functions. Each committee reviews and comments on draft technical information and reports. In addition, the PAC can set some priorities for RAP development and identify concerns to be addressed. The PAC does not supersede the role of the RAP team; rather, the two committees cooperate to address matters from a broad perspective. In this way, recommendations forwarded to the Parties of the 1987 Great Lakes Water Quality Agreement are a joint effort and represent a greater range of aspirations and expectations.

Finally, there was substantial public input to the *Time to Decide* discussion paper and many action recommendations were endorsed by the larger stakeholder group. Input was received from a range of groups and organizations, including a Grade Three environmental studies class, municipal officials, local Conservation Authorities, developers, sports fishermen and farmers who use alternative and ecologically-sound farming practices.

The Quinte RAP education and consultation effort has made advances as well. Public awareness of the local water quality problems has increased, and formal teaching programs include many Bay of Quinte RAP materials. A coordinated Bay of Quinte teaching package also has been requested. The communication links have generated a more environmentally-responsible Bay of Quinte populace and other environmental services (e.g. solid waste reduction, recycling, and household hazardous waste programs) are being studied.

As an additional measure to ensure ownership of the Bay of Quinte RAP throughout the region and among all segments of the local society, the RAP team and the PAC jointly prepared a discussion paper. The report was written for a broad audience and was intended to promote informed discussion, decisionmaking, and consensus building.

The report was based on the historical Project Quinte data base and the technical information prepared by the RAP team. Four specific water quality problems were summarized: (1) nutrient enrichment, (2) habitat destruction, (3) bacteriological contamination and (4) toxic contaminants. Numeric restoration objectives were defined only for the nutrient enrichment issue, and general objectives focusing on human health concerns and protection of wetlands were suggested for other water quality concerns.

Realistically, ecosystem and water quality conditions in the Bay of Quinte cannot be returned to a presettlement or pristine status.

The objectives proposed in the *Time to Decide* report, however, were designed to recreate a trophic status similar to conditions in the 1930s. Restoration objectives for nutrient enrichment include: (1) reduce the average concentration of total phosphorus in the water column from 47 μg L^{-1} to 30 μg L^{-1} (2) reduce average algal density from 8.2 mm^3 L^{-1} to 4.0—5.0 mm^3 L^{-1}; and (3) increase the potential area of macrophyte coverage from an estimated 38 km^2 to 45 km^2 (15 to 17 square miles). These objectives were selected, in part, because each could be numerically supported by the outcome of the Phosphorus-Ecosystem Model. Additional numeric objectives were not established because of reduced confidence in the model's predictions.

Throughout the consultation period, the PAC has pressed for, and strongly endorsed protection of the remaining natural shoreline, fish habitat and wetland areas in and around the Bay of Quinte. However, physical habitat destruction is often irreversible, depending on the nature of the destruction. Thus, the resolution adopted by PAC and presented in the *Time to Decide* report was to restore damaged habitat wherever practical and apply a "no net loss" habitat and wetland policy.

Bacteriological contamination restoration objectives aim to eliminate human health hazards and improve recreational uses, including zero beach closures in the Area of Concern and maintaining fecal coliform densities of less than 100 organisms per 100 mL of water at any location within the bay. The potential impacts of persistent toxic contaminants on human health are less well understood because the data base is incomplete and the movement of contaminants through the food chain is poorly understood. The report's stated objective is to eliminate all inputs of persistent toxic contaminants from all sources within the Area of Concern.

The *Time to Decide* report was widely advertised, mailed to all stakeholders and made available to other interested parties. The RAP team and the PAC also jointly provided technical discussions, presentations, and information exchanges. Comments on the proposed course of remedial action were received, evaluated and formalized by the PAC, and this information will form the heart of the Bay of Quinte RAP.

The consultation actions provided the public with an understanding of the technical aspects of the Bay of Quinte RAP. Through the PAC, the public played a direct role in the plan's development. This generated a sense of public ownership and local control, and initiated discussions regarding ongoing public involvement, the distribution of responsibilities, and a structure for a more permanent Quinte RAP public-agency relationship. However, the expectations and momentum raised in the technical review and public consultation phase must be carried into the next stages of negotiation, commitment, and implementation if the sharing of power, responsibility, and control over the Bay of Qunite's environmental destiny is to be realized.

RAP Challenges/Implementation

Canada's commitment to develop RAPs for 17 Areas of Concern was made with the 1987 Protocol to the Great Lakes Water Quality Agreement and the Canada-Ontario Agreement (COA) provided the mechanism for the province of Ontario's commitment to Canadian

RAP development. Via these agreements, a RAP Steering Committee and 17 RAP teams were established. The steering committee provides planning, direction, communication and funding, while RAP teams develop the site-specific RAPs. The Bay of Quinte RAP team includes two federal and three provincial agencies and is assisted by an enthusiastic and capable Public Advisory Committee. In most cases, this simple framework has allowed for orderly RAP development.

The existing RAP framework, however, does not clearly provide the means to implement the plan. Likewise, the mechanisms to apply sustainable development and an ecosystem approach have not become part of either the RAP program or the Canada-Ontario Agreement (COA).

The ecosystem approach requires the relationship of all watershed activities to be considered in water quality impairment. To synchronize RAP development, all research and activities directly affecting water quality in the Area of Concern should be evaluated and priorized by the RAP team. In addition, funding for RAP activities should be coordinated by the RAP Steering Committee with advice from the RAP teams.

RAP implementation may require an additional framework and/ or agreement. The Bay of Quinte RAP is considering either a legal framework and/or an institutional framework arising from the existing Great Lakes Water Quality and Canada-Ontario Agreements. The latter would amend COA to create a RAP implementation framework that specifies roles, responsibilities, funding, reporting, and monitoring mechanisms, as well as provide for the development of a site-specific Bay of Quinte RAP Implementation Agreement. Although the tiered agreement lacks the perceived assurances of a legal framework, it is a reasonable compromise, it would maintain momentum and would establish the links of accountability between the Quinte RAP through COA to the signatories of the Great Lakes Water Quality Agreement.

The Great Lakes Water Quality Agreement emphasizes the importance of integrating all interactive elements (chemical, physical and biotic) within a total Great Lakes basin framework to ensure that lasting solutions to our environmental problems are designed and implemented (International Joint Commission 1978; Vallentyne and Beeton, 1988). Moreover, in defining the terms of reference for the development of RAPs for the 43 Areas of Concern, the International Joint Commission encouraged the application of an ecosystem approach (Hartig and Vallentyne, 1989). To successfully apply this ecosystem approach, a RAP must include more information on the autecology (i.e. species ecology) of the main components of the ecosystem than did older approaches (e.g. "egosystemic" and "piecemeal" approaches described by Christie et al. (1986)). An additional stipulation of the RAP process was for the integration of environmental, social, economic and political components. Thus, the RAP would be "owned"—and subsequently implemented—by all segments of society.

The Bay of Quinte RAP Team has benefited from years of research and monitoring via Project Quinte (Minns et al. 1986a) prior to the RAP process. Considerable understanding of the important physical and chemical characteristics, as well as the major biotic components (phytoplankton, macrophytes, zooplankton, benthos, and fish communities), was gained.

Consistent with an ecosystem approach, some attempts to model the interactions of several of these components were made (Johnson 1986; Hurley et al. 1986; Minns et al. 1987; Nicholls and Hurley, 1989). A dramatic example involves the ecosystem linkages believed to exist between fish, macrophytes and algal communities. It is hypothesized that small corrective measures on one component would cause "desirable" changes in other ecosystem structures and functions. Ecosystem health and stability, and hence human use potential, would improve "naturally" through a snowballing effect.

As a means of illustration, it is believed that the 1980s fish community, dominated by walleye, is less stable than a community which would include representation of several other warmwater piscivores (e.g. northern pike, largemouth bass, longnose gar, and bowfin). The more diverse community would be expected to exert greater control on zooplanktivorous and benthivorous species such as alewife and white perch. A decline in these less desirable, non-native species should lead to higher densities of planktonic and benthic herbivores and, in turn, contribute to lower densities of algae in the Bay of Quinte. This new community structure may result in improvements in the underwater light climate and recolonization of macrophytes, and therefore, enhancement of habitat for the warmwater piscivore community. In this way, the cycle of

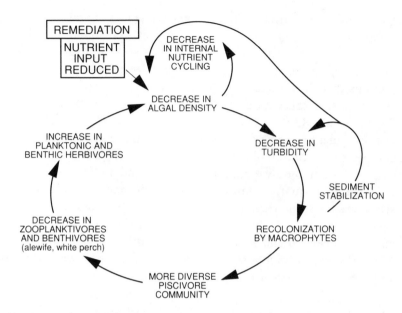

Fig. 15. Illustration of remediation effects on biotic interactions in the Bay of Quinte. Remediation to some threshold level (not necessarily complete elimination of the perturbation) causes changes in community structure and function which lead to further enhancement of ecosystem stability.

water quality improvements and aquatic ecosystem enhancement continues (Figure 15). At the same time, the ecosystem approach may have detractions. Employing the above hypothesis, the thriving walleye sport fishery might be negatively affected—increasing northern pike might reduce walleye abundance. Also, as alewife numbers are controlled, in part, by walleye predation, alteration of the current species mix may allow alewife numbers to increase again. In other words, the course of action is not simple and decisions must be made wisely.

Conclusions

The Bay of Quinte RAP process is moving forward and giving a more prominent role to an integrated ecosystem approach. The initiative is addressing environmental concerns and solving problems by including essential human elements (e.g. societal goals, economic development, and implications of continued population growth) in the management formula. The stakeholders of the Bay of Quinte, for example, rejected the "flushing" option on the grounds that flow augmentation and dilution can no longer be considered a societal solution under the ecosystem framework of the Bay of Quinte RAP.

It is evident from past scientific studies in the Bay of Quinte that complete elimination of anthropogenic nutrient loadings is not necessary. Rather, ecosystem rehabilitation and increased enhancement of the aquatic community structure and function will be stimulated if some threshold level of reduction is achieved. Nonetheless, the Bay of Quinte RAP process has shown that the human element is central and essential. Some sacrifices of human use will be necessary to achieve any long-term benefits associated with ecosystem stability (Table 14).

TABLE 14.   The Major Characteristics and Human Implications of a Perturbed and a Restored Bay of Quinte Ecosystem (nutrient loading–trophic state example)

Perturbed State (excessive nutrient inputs)	Human Implications
Algal blooms	aesthetic and water treatment problems (including costs)
Turbid water	
Few macrophytes	
Low diversity in fish community ("boom and bust" cycles)	alternating periods of "good" and "poor" fish harvests (restricted to few species and largely unpredictable)

Remediation Implemented (reduced nutrient inputs)	Human Implications
Improved water treatment	costs of remediation
Improved water quality and clarity	more attractive for recreational use and less troublesome as a source of domestic water supply
More macrophytes	problems of access and boat operation
More diverse and stable fish community	more consistent long-term yields of fish (plus more species)

REFERENCES

Bay of Quinte RAP Coordinating Committee and Bay of Quinte RAP Public Advisory Committee. 1989. *Time to Decide: A Discussion Paper. Bay of Quinte RAP Technical Report.* Kingston, Ontario.
Bowlby, J.N., A. Mathers, D.A. Hurley, and T.H. Eckert. 1990. *The resurgence of walleye in Lake Ontario.* Great Lakes Fishery Commission, Walleye Rehabilitation Workshop. Ann Arbor, Michigan.
Canada and the United States. 1987. *Protocol to the 1978 Great Lakes Water Quality Agreement.* Windsor, Ontario, Canada.
Christie, W.J., M. Becker, J.W. Cowden and J.R. Vallentyne. 1986. Managing the Great Lakes as a home. *J. Great Lakes Res.* 12:2–17.
Crowder A. and M. Bristow. 1986. Aquatic macrophytes in the Bay of Quinte, 1972–82. *In* C.K. Minns, D.A. Hurley and K.H. Nicholls (ed.) Project Quinte: Point-source phosphorus control and ecosystem response in the Bay of Quinte, Lake Ontario. *Can. Spec. Pub. Fish. Aquat. Sci.* 86:114–127.
Ecologistics Limited. unpubl. *Bay of Quinte Remedial Action Plan Socio-Economic Assessment of Proposed Remedial Measures.* Policy and Planning Branch, Environment Ontario, Toronto, Ontario, Canada.
ESSA Limited. 1988. *An Ecosystem Based Analysis of Remedial Options.* Bay of Quinte Remedial Action Plan Coordinating Committee. Technical Report #6. Kingston, Ontario, Canada.
Hartig, J.H. and J.R. Vallentyne. 1989. Use of an ecosystem approach to restore degraded areas of the Great Lakes. *AMBIO* 18:423–428.
Holling, C.S. (ed.). 1978. *Adaptive Environmental Assessment and Management.* John Wiley and Sons, Toronto.
Hurley, D.A. and W.J. Christie. 1977. Depreciation of the warmwater fish community in the Bay of Quinte, Lake Ontario. *J. Fish. Res. Board Can.* 34:1849–1860.
Hurley, D.A., W.J. Christie, C.K. Minns, E.S. Millard, J.M. Cooley, M.G. Johnson, K.H. Nicholls, G.W. Robinson, G.E. Owen, P.G. Sly, W.T. Geiling and A.A. Crowder. 1986. Trophic structure and interactions in the Bay of Quinte, Lake Ontario, before and after point-source phosphorus control. *In* C.K. Minns, D.A. Hurley and K.H. Nicholls (ed.). Project Quinte: Point-source phosphorus control and ecosystem responses in the Bay of Quinte, Lake Ontario. *Can. Spec. Publ. Fish. Aquat. Sci.* 86:259–270.
International Joint Commission. 1978. *The ecosystem approach.* Great Lakes Science Advisory Board, Windsor, Ontario, Canada.
Johnson, M.G. 1986. Phosphorus loadings and environmental quality in the Bay of Quinte, Lake Ontario. *In* C.K. Minns, D.A. Hurley and K.H. Nicholls (ed.) Project Quinte: Point-source phosphorus control and ecosystem response in the Bay of Quinte, Lake Ontario. *Can. Spec. Publ. Fish. Aquat. Sci.* 86:247–258.
Johnson, M.G. and O.C. McNeil. 1986. Changes in abundance and species composition in benthic macroinvertebrate communities of the Bay of Quinte, 1966–84. *In* C.K. Minns, D.A. Hurley and K.H. Nicholls (ed.) Project Quinte: Point-source phosphorus control and ecosystem response in the Bay of Quinte, Lake Ontario. *Can. Spec. Pub. Fish. Aquat. Sci.* 86:177–189.

Johnson, M.G. and G.E. Owen. 1971. Nutrients and the budgets in the Bay of Quinte, Lake Ontario. *J. Water Pollut. Control Fed.* 43:836–853.

McCombie, A.M. 1967. A recent study of phytoplankton of the Bay of Quinte 1963–1964. *Proc. Tenth Conf. Great Lakes Res.* pp. 37–62.

Michalski, M. 1987. Bay of Quinte Remedial Action Plan: Progress Report, February 1987. *Bay of Quinte RAP Coordinating Committee*, Kingston, Ontario, Canada.

Millard, E.S. and M.J. Johnson. 1986. Effect of decreased phosphorus loading on primary production, chlorophyll *a* and light extinction in the Bay of Quinte, Lake Ontario, 1972–1982. *In* C.K. Minns, D.A. Hurley and K.H. Nicholls (ed.) Project Quinte: Point-source phosphorus control and ecosystem response in the Bay of Quinte, Lake Ontario. *Can. Spec. Publ. Fish. Aquat. Sci.* 86:270 pp.

Minns, C.K., D.A. Hurley and K.H. Nicholls. 1986a. Project Quinte: Point-source phosphorus control and ecosystem response in the Bay of Quinte, Lake Ontario. *Can. Spec. Publ. Fish. Aquat. Sci.* 86:270 pp.

Minns, C.K., G.E. Owen and M.J. Johnson. 1986b. Nutrient loads and budgets in the Bay of Quinte, Lake Ontario. *In* C.K. Minns, D.A. Hurley and K.H. Nicholls (ed.) Project Quinte: Point-source phosphorus control and ecosystem response in the Bay of Quinte, Lake Ontario. *Can. Spec. Publ. Fish. Aquat. Sci.* 86:270 pp.

Minns, C.K., E.S. Millard, J.M. Cooley, M.G. Johnson, D.A. Hurley, K.H. Nicholls, G.W. Robinson, G.E. Owen and A. Crowder. 1987. Production and biomass size spectra in the Bay of Quinte, an eutrophic ecosystem. *Can. J. Fish. Aquat. Sci.* 44 (Suppl. 2):148–155.

Mudroch, A. and J.A. Capobianco. 1980. Impact of past mining activities on aquatic sediments in Moira Basin, Ontario. *J. Great Lakes Res.* 6:121–128.

Nicholls, K.H., L. Heintsch, E. Carney, J. Beaver and D. Middleton. 1986. Some effects of phosphorus loading reductions on phytoplankton in the Bay of Quinte, Lake Ontario. *In* C.K. Minns, D.A. Hurley and K.H. Nicholls (ed.) Project Quinte: Point-source phosphorus control and ecosystem response in the Bay of Quinte, Lake Ontario. *Can. Spec. Publ. Fish. Aquat. Sci.* 86:270 pp.

Nicholls, K.H. and D.A. Hurley. 1989. Recent changes in the phytoplankton of the Bay of Quinte, Lake Ontario: the relative importance of fish, nutrients, and other factors. *Can. J. Fish. Aquat. Sci.* 46:770–779.

Ontario Ministry of Natural Resources. 1989. *Lake Ontario Fisheries Unit 1988 Annual Report.* Toronto, Ontario, Canada.

Poulton, D.J. 1989. Bay of Quinte Remedial Action Plan: 1987 Bacteriological Water Quality at Trenton, Deseronto and Picton, Bay of Quinte, *Bay of Quinte RAP Technical Report* No. 5. Bay of Quinte RAP Coordinating Committee, Toronto, Ontario, Canada.

Project Quinte. unpubl. *Project Quinte Annual Report 1988.* Toronto, Ontario, Canada.

Ridgeway, M.S., D.A. Hurley, and K.A. Scott. 1990. The impacts of winter temperature and predation on the abundance on alewife (*Alosa pseudoharengus*) in the Bay of Quinte, Lake Ontario. *J. Great Lakes Res.* 16:11–20.

Sly, P. 1986. Review of postglacial environmental changes and cultural impacts in the Bay of Quinte. *In* C.K. Minns, D.A. Hurley and K.H. Nicholls (ed.) Project

Quinte: Point-source phosphorus control and ecosystem response in the Bay of Quinte, Lake Ontario. *Can. Spec. Publ. Fish. Aquat. Sci.* 86:270 pp.

Vallentyne, J.R. and A.M. Beeton. 1988. The ecosystem approach to managing human uses and abuse of natural resources in the Great Lakes basin. *Environ. Conserv.* 15:58–62.

Chapter 9

Rochester Embayment's Water Quality Management Process and Progress, 1887–1990

Margaret E. Peet, Richard Burton, Richard Elliott, Philip Steinfeldt, John Davis, Sean Murphy, and Andrew Wheatcraft

> "Lake Ontario is a key to sustaining our community—as a source of drinking water and industrial water, and in terms of recreational opportunities and quality of living here in Monroe County. I also believe that freshwater is our secret weapon in the northeast. Areas of the south and west already compete for a limited supply of clean, fresh water. In the northeast, we're blessed with an abundance of fresh water, and we must take this opportunity to improve and protect this birthright. Our remedial action plan will likely call for everyone—citizens, government and industry—to take actions to improve our water. Our challenge is to effectively communicate our remaining needs and enlist the help of all sectors of our community."
>
> Thomas R. Frey
> County Executive
> Monroe County, New York

Introduction

The Rochester Embayment on Lake Ontario is one of 43 Areas of Concern identified in the Great Lakes basin where a remedial action plan (RAP) is required to restore impaired uses. This is being accomplished as a cooperative effort led by the County of Monroe and the State of New York. Remedial action planning is not new in the area; water quality problems as far back as 1887 have sparked community actions in Monroe County, where the City of Rochester serves as the urban center.

Monroe County is located near the midpoint of the south shore of Lake Ontario (Figure 16). The Genesee River, with a 6,420 km² (2,480 square mile) drainage basin, extends south into the State of Pennsylvania and flows into the Rochester Embayment of Lake On-

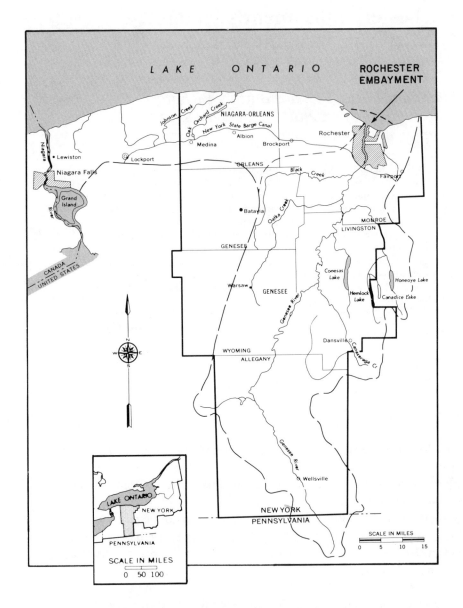

Fig. 16. Monroe County, New York, watersheds addressed in the
Rochester Embayment Remedial Action Plan

tario along the Monroe County shoreline. The county's urban center is located approximately 11.3 km (7 miles) upstream from the mouth of the Genesee River. Approximately 6.4 km (4 miles) to the east of the Genesee River mouth is the 6.7 km (4.2 mile) long Irondequoit Bay, a 395 km² (153 square mile) watershed. The remainder of the county is drained by streams that flow directly into the embayment.

The purpose of this chapter is to provide a historical overview of water pollution control efforts for the Rochester Embayment RAP as the latest step in the evolution of water quality management planning. Emphasis is placed on multi-media watershed planning consistent with the ecosystem approach called for in the Great Lakes Water Quality Agreement (United States and Canada, 1987).

Background

City of Rochester Sewage Disposal 1887 to 1960

Rochester has been a pioneer in many respects in dealing with sewerage and sewage disposal problems (Metcalf and Eddy Engineers, 1929). Until about 1890, sewer systems discharged at various points into the Genesee River. The city's population was small and discharges did not produce visually objectionable conditions. However, health problems due to water quality periodically occurred, including outbreaks of waterborne cholera in 1832, 1849 and 1852. These problems led to construction of sewers and a centralized water supply (McKelvey 1956).

The gradual degradation of water quality in the Genesee River and Lake Ontario led to a series of studies in 1887, 1907 and 1929. During this time the area experienced dramatic population growth and a resulting concern for intermunicipal water pollution. Three studies considered the need for a comprehensive sewer system to link all municipalities, upgrade and expand existing sewage treatment, and provide overflow structures to discharge combined sanitary/storm waters during wet weather. The recommendations of these studies recognized the need for 1) a comprehensive sewage interceptor system to link discharges in the City of Rochester and the surrounding areas into a central wastewater treatment plant; 2) continued tracking of combined sewer overflows and the eventual need to identify alternatives to eliminate these overflow structures; and 3) expansion and upgrading of treatment facilities, including more efficient hydraulic systems, and possible chlorination to reduce bacteria levels (Metcalf and Eddy Engineers, 1929; Kuichling 1907; Burton 1975; and Day 1967).

1960s: Suburban Growth, Comprehensive Sewer
Planning and Citizen Involvement

Considerable community and governmental concern grew for water quality in the 1960s. This community concern coincided with a period of substantial growth in population and new housing. Much of this post–World War II growth occurred in the suburbs, where wastewater was treated and directed to smaller streams rather than to the lake. Something had to be done to alleviate the stress on Irondequoit Creek in particular, a tributary of Rochester Embayment.

In 1964, Monroe County initiated a comprehensive sewerage study. While this study was being completed, local citizens showed their concern for water quality by documenting community water quality concerns, specifically chemical oxygen demand and bacteria levels (Rochester Academy of Science 1982).

In 1965, the $1 billion New York State Pure Waters Bond Act was approved, which increased state aid to construct sewage treatment plants, reimbursed costs for treatment plant operation and maintenance, and enforced the state's laws against water pollution (Day 1967).

Despite improvements in wastewater treatment plants by the late 1960s, existing sewerage systems in parts of the county became a major constraint on continuing development (Christensen 1968). As the situation became desperate, the Monroe County Pure Waters Agency (MCPWA) was established to address the problems of inadequate sewerage facilities in the county (Christensen 1968). Most streams had serious water quality problems, and it was concluded that a countywide interceptor system was needed to carry the sewage flow to regional treatment facilities and to abandon most town and village plants (Christensen 1968).

The 1969 Pure Waters Master Plan stemmed from the combined efforts of the Monroe County Comprehensive Sewerage Study and the Monroe County Pure Waters Agency. The plan reported severe oxygen depletion along several portions of the Genesee River as a result of demand from food processing wastes, municipal and industrial primary effluent, and combined sewer overflows in the City of Rochester. Relatively small streams such as the Genesee River were suffering from similar conditions (Monroe County Pure Waters 1969). By 1969, Lake Ontario was suffering from cultural eutrophication, and nuisance *Cladophora* growth on shorelines and alewife dieoffs were common. Because of the discharge of untreated and partially treated sewage and the fact that effluent plumes hug the shores under most wind/current conditions, Ontario beach was closed completely because coliform densities exceeded those permitted by law (Burton 1975).

The 1969 Pure Waters Master Plan

In 1969, a Pure Waters Master Plan was developed which summarized conditions of existing sewage facilities and receiving waters in Monroe County and outlined a plan of action to handle municipal sewage and improve water quality. Its general policies were directed at five areas:

- *Receiving Waters:* Only Lake Ontario and the Genesee River were recognized as having the capacity to receive wastewater effluents, and the elimination of effluent into the Genesee was established as a long-term goal.
- *Treatment Facilities:* All facilities were to provide 85 percent removal of biochemical oxygen demand and suspended solids, provide 80 percent removal of phosphorus, provide year-round disinfection of effluent, and encourage placement of outfalls beyond the bounds of the Rochester Embayment.
- *Sewer Collection System:* The maximum possible volume of combined sewer overflows was to be transported to treatment facilities, with the remaining to be chlorinated prior to discharge. Any new development was to be built with separate storm and sanitary sewers.
- *Industrial Wastewater:* Except for coolants and process water relatively free of pollutants, industries were encouraged to discharge wastewater into public sewage systems.

- *Others:* Septic tanks were permitted where sewers were not available, and stricter actions were to be taken to reduce urban and agricultural erosion.

Implementation of Countywide Pure Waters Plan: 1970 to 1990

The heart of the Pure Waters Plan (Monroe County Pure Waters Agency 1969) called for creating four regional sewerage systems based on drainage basins, each with a wastewater treatment facility and a network of interceptors. Implementation of major portions of the master plan resulted in a comprehensive sewer system being built, and residents in those districts were now able to see facilities that were planned for their benefit. As a result, support for this program developed quickly.

For the most part, construction of the county operated system was completed by 1978 (Quinn 1979) and includes three secondary wastewater treatment facilities, two of which discharge effluent to Lake Ontario and one that discharges to the Genesee River. Sludge from the three county plants is dewatered and incinerated on site. The construction of 160 km (100 miles) of interceptor sewers phased out 28 small treatment plants. In addition to the five county operated systems, six town and village collection and treatment systems still operate and discharge to either Lake Ontario, the Genesee River, its tributaries or Northrup Creek (Monroe County Water Quality Management Agency 1987).

The cost to intercept suburban sewage treatment plants was $218.4 million. Funds were provided from federal ($102.4 million), state ($62.7 million) and local level ($53.3 million). Suburban households pay about $55 per year (Davis 1989) to support the treatment program.

Combined Sewer Overflow Abatement Program

The combined sewer overflow problem in the City of Rochester has been attacked in three ways, beginning in the 1970s and continuing in the 1990s.

1. The Combined Sewer Overflow Abatement Program (CSOAP) uses several large underground storage/conveyance tunnels

constructed deep beneath the city. During storms, combined sewage from smaller collector sewers passes through regulatory stations and is temporarily stored in these large underground storage tunnels. After the storms, and as capacity is available at the upgraded Van Lare Wastewater Plant, the combined sewage is conveyed to the wastewater treatment plant.

The CSOAP tunnel construction includes 40 km (25 miles) of tunnel bored beneath the city, which has been divided into 14 separate construction projects. As of 1990, nine tunnels are completed and operating and two are under construction. Implementation of the master plan in the city, including the CSOAP program, and improvements and odor abatement at the Van Lare Plant have cost $742.6 million—51 percent of which has been provided by the federal government, 13 percent from the state, and 36 percent from the local district (Davis 1989). During 1989, 9.46 million m^3 (2.5 billion gallons) of combined sewage that would previously have been discharged to waterways untreated, or, in some cases backed up into basements, was conveyed to the Van Lare Wastewater Treatment Plant via the tunnel system.

A mathematical model of the existing Rochester sewer system was developed to provide information to design the new system. Overloaded sections of sewers were identified and improvements to the system were chosen based on cost as well as social, political and environmental factors. Mathematical and physical models of the proposed tunnels were then developed to refine the design and key components of the tunnel system (Joint Venture 1977).

2. The Van Lare Wastewater Plant was upgraded to accept the new volume of wastewater now stored in tunnels.

3. Best management practices were developed to maximize the capacity of the existing surface sewer system.

Industrial Wastewater Treatment

In 1990, industries can discharge either to a municipal conveyance system, in which case the effluent is treated at a municipal sewage treatment facility, or treat the flow and discharge it directly into waterways. In the first case, the effluent must be pretreated to comply with federal and state pretreatment requirements that were promulgated by the federal government in 1978 to protect municipal treatment plants from substances that would influence the treatment processes or exceed discharge limits. Since 1972, the Monroe County Department of Pure Waters has conducted a rigorous industrial discharge permit program under the authority of the county's Sewer Use Law for industries discharging their wastewater to the county's sewage system (Monroe County Water Quality Management Agency 1987). Industrial wastewater is sampled by the discharger, the county and the New York State Department of Environmental Conservation (NYSDEC). The county also samples the collection system and three wastewater treatment plants on a quarterly basis for priority pollutants.

Sixty industrial facilities in Monroe County have State Pollution Discharge Elimination System (SPDES) permits to treat the flow and discharge wastewater directly into waterways. Forty of these facilities represent significant discharges to county waters. Some industries have improved their wastewater treatment operations, such as Kodak and Rochester Gas and Electric Corporation.

Kodak

Kodak Park has over 200 major manufacturing buildings and lies on a bluff about 45.7 m (150 feet) above the Genesee River in the City of Rochester. Prior to introduction of regulatory requirements, Kodak had a small screening facility for wastewater and had installed and operated a number of at-source solvent and metals recovery operations to treat its waste. In 1957, a primary treatment plant was built on the banks of the Genesee River to handle 1.1 m^3/second (25 million gallons per day) (Lindsley 1974). Studies in the 1960s indicated that an effluent with high biochemical oxygen demand and heavy metal levels was still being discharged from the site (Day 1973). In 1970, a secondary biological treatment plant was put into operation (Lindsley 1974).

The 1970 plant upgrade removed in excess of 90 percent of the suspended solids and biochemical oxygen demand loading and about 75 percent of the heavy metals (Day 1973). During the 1970s and

1980s, Kodak examined advanced treatment technologies and worked with manufacturing personnel to reduce the biochemical oxygen demand load at source, which has resulted in an average biochemical oxygen demand (exerted in 28 days) discharge well below the SPDES permit level of 9,526 kg (21,000 pounds) per day.

Kodak reported and followed up on periodic exceedances in SPDES permit levels during 1989. Working with state and local governments, the company has recently agreed to have an environmental audit conducted to address concerns about chemical spills and permit level exceedances (NYSDEC 1990).

Rochester Gas and Electric Corporation

Rochester Gas and Electric Corporation (RG&E) operates a 75 megawatt, coal-fired, steam electric generating station (Beebee Station) next to the Genesee River, between the upper and middle falls in the City of Rochester. This facility is one of several power generating plants they operate in the area. The U.S. Army Corps of Engineers in 1968 reported that RG&E discharged fly ash to the river, and the corporation now has three permitted water discharges at its Beebee Station.

RG&E assessed the impact of the thermal discharges on the aquatic community in 1977 and generally found that the indigenous aquatic community in that portion of the river was protected and was propagating (RG&E 1977). Another RG&E study conducted from 1975 to 1985 evaluated numbers and species of fish collected in the main circulation water travelling screens or passing through the plant, and found that, in general, impingement rates fluctuate greatly from year to year with relatively high rates for alewife, gizzard shad, and smelt during fall and early winter (RG&E 1985).

RG&E operates a central treatment facility that collects and treats flows from all potential pollutant sources within this coal-fired plant. The quality of the discharge water from this plant is below ambient river levels 90 percent of the time for total suspended solids. In addition, these discharges are equal to or below river ambient levels for metals greater than 50 percent of the time.

Onsite Sewage Disposal

Prior to the formation of the Monroe County Department of Health in 1958, onsite sewage disposal systems was regulated by the New York State Department of Health in conjunction with inspection work done by towns. County and state staff did percolation tests and deep hole inspections, as well as site inspections. The state reviewed plans and recommended approval or disapproval. With the formation of the County Department of Health in 1958, subdivision approval became the responsibility of the county. The County Health Department worked with municipal building inspectors in an installation inspection program.

By 1973, the county was reviewing, approving and making inspections for all but one town. During the early 1970s, the county participated in a Genesee-Finger Lakes Health Planning Council effort to develop a model, residential, onsite sewage disposal code. In 1974, Monroe County adopted Article IIA of the Monroe County Sanitary Code patterned after the model and in 1976, the New York State Department of Health promulgated a uniform set of residential onsite sewage disposal standards, which the county began using with an addendum recognizing problems and practices unique to Monroe County.

Dredging

The U.S Army Corps of Engineers (Corps) has been authorized to regularly dredge the Genesee River at the Rochester Harbor since 1829 under the River and Harbor Acts. The Corps is authorized to dredge a 6.4–7.3 m (21–24 feet) deep channel from Lake Ontario to a turning basin located 3.1 km (1.9 miles) upstream. In the 1960s, approximately 252,300 m³ (330,000 cubic yards) were dredged from the river (U.S. Army Corps of Engineers 1968), and the impact of the dredging operation on water quality has been an issue since that time. The two major concerns about Genesee River dredging are the open lake disposal site located approximately 2.4 km (1.5 miles) offshore and the potential impacts of the dredging itself on the water quality in the river and nearby beaches.

In 1969, the Corps published a report on dredging and water quality problems in the Great Lakes to identify the need for alternative methods of dredged spoil disposal. The report found that the best alternative disposal method would be diked containment areas near the dredging projects and recommended that legislation be enacted to accomplish this. A preliminary study evaluated seven shoreline sites for diked disposal to hold 10 years' worth of dredged materials.

All recommended sites received stiff opposition from the State Conservation Department, the City Parks Department, or others (U.S. Army Corps of Engineers 1968). Before establishing disposal facilities, the Corps was directed to obtain concurrence of appropriate local governments, consider the views and recommendations of the U.S. Environmental Protection Agency, and comply with the U.S. Environmental Protection Act of 1969 and the Rivers and Harbors Act of 1970. However, a year after the final Environmental Impact Statement was submitted, the State of New York refused to allow continued open lake disposal and the county passed a resolution opposing a land disposal site, supporting interagency cooperation and an effort to identify and abate the pollutant sources. The issue soon turned from whether the Corps would dredge, to a discussion of the chemical characteristics of the sediment (particularly metals).

In 1974, in an effort to deal with the continuing need for a disposal site, the county legislature asked the Corps to design a diked disposal area at a site between Irondequoit Bay and the Genesee River mouth. This effort was halted by the Corps and U.S. EPA because methods to evaluate polluted sediment were changing, there was a lack of information on the impacts of open lake disposal, progress was occurring in other remedial programs for the improvement of water and sediment quality, and concern was expressed for the economic viability of the Port of Rochester (U.S. EPA 1975). It was concluded that the best and most reasonable approach to take was to continue to aggressively pursue programs to reduce the active sources of municipal and industrial pollution to the sediments of the lower Genesee River (Burton 1984).

The other concern about water quality stems from the use of hopper dredges. The problems associated with hopper dredges are minimal, unless the dredge is operated in an overflow method. With the overflow method, the hoppers are filled beyond volumetric capacity until the density of the load reaches a level that has been predetermined to give maximum operational efficiency. The overflow returns the less dense material to the river, causing considerable local turbidity. This method also allows toxic pollutants and bacteria to re-enter the river, resulting in a marked degradation in the water quality both from the standpoint of appearance and level of pollutants (Monroe County EMC 1983).

Monroe County was concerned as early as 1975 about the impacts of overflow dredging on water quality. The county first requested that overflow dredging be restricted and then eliminated in 1977 to reduce the impacts of dredging on water quality. A subsequent hopper dredge in 1982 again caused concern over water quality, and after extensive monitoring in 1986, it was decided that no overflow dredging would be conducted in the Rochester Harbor. All subsequent dredging has been by clamshell or hopper dredge without overflow.

Lake Ontario Beaches

Information from as far back as 1912 indicates that currents and winds transport Genesee River bacteria to Lake Ontario beaches. Data from 1929 indicate that lakeshore beach water quality was marginal to unacceptable by modern standards. Records of lakeshore water quality just west of Ontario Beach show a gradual increase in fecal coliform levels from 1932 to the mid-'50s with a slight decrease through 1962. The decline in coliforms in the late 1950s was likely due to the improvements in sewage collection and waste treatment in the Town of Greece and to a lesser extent in the City of Rochester (Burton 1975).

Prior to 1967, the Monroe County Department of Health granted permits for public bathing beaches at Ontario Beach Park based on the absence of public health hazard evidence and reluctance to deny recreational access. In 1967, the State Public Health Law set fecal coliform bacteria limits for bathing water. Because of violations of this standard, the New York State Health Department recommended Ontario Beach closure in March 1967.

In 1972 the Monroe County Department of Health, in cooperation with the University of Rochester, began an intensive water quality sampling program beginning at Ontario Beach and expanding to Durand Eastman Beach. This program was expanded to include several other lakeshore areas by 1975, and identified three major sources of coliforms to Ontario Beach: 1) stormwater runoff from drainage basins west of the Genesee River; 2) Genesee River combined sewer overflows and stormwater runoff; and 3) decomposing organic material (including *Cladophora*) along the beach shore (Burton 1975).

The 1975 study recommended a conditional operating permit for the bathing beach at Ontario Beach Park, recognizing that an occasional impairment of water quality from stormwater and combined sewer overflows could be anticipated. It also recommended that *Cladophora* be promptly removed from beaches. A monitoring program continues to verify or modify closure criteria, and the conditional permit program instituted in 1976 continues today. Criteria were updated in 1989 to further specify operating guidelines and subsequent actions.

The 1980s: Focus on Nonpoint Sources of Pollution

By 1980, the efforts of the Pure Waters program resulted in the elimination of nearly all sanitary sewage effluent to Irondequoit Bay and its tributary streams. Despite this, Irondequoit Bay remained in a hypereutrophic state (Figure 17). Considerable investment has been made in the portion of the Pure Waters program to divert sewage from this basin because residents and community leaders viewed the bay as a valuable recreational and aesthetic resource that deserved restoration.

Research

Impacts of Urban Stormwater Runoff
During the 1970s, U.S. EPA sponsored the Nationwide Urban Runoff Program (NURP) to define the sources, transport, and accumulation patterns of some regional stormwater pollutants, document the effects of this pollution and identify remedial measures. In 1979, U.S. EPA, NYSDEC and Monroe County agreed that the Irondequoit basin would be one of 28 regional study areas. To build on this effort, the U.S. Geological Survey (U.S. GS) and Monroe County entered into a cooperative agreement to assist in assessing urban runoff quantity and quality. There were five major conclusions from this study:

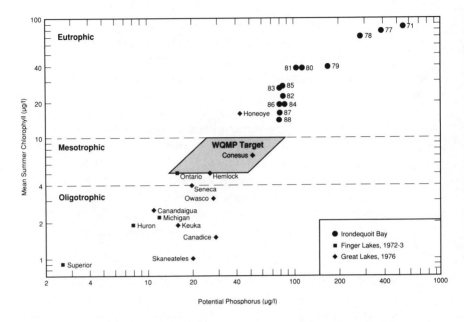

Fig. 17. Irondequoit Bay changes in trophic state

- Phosphorus is the most easily controlled nutrient contributing to the present eutrophic state of the bay.
- Decreases in phosphorus in the bay can be predicted from reductions in stormwater phosphorus loading to the bay.
- The largest amount of runoff phosphorus per unit of land area is from urban areas of the basin.
- The most cost effective control measure to reduce runoff phosphorus from Irondequoit Creek involves the use of the wetland immediately upstream of the bay as a natural "treatment" area.
- Combinations of other control measures are needed, including instream impoundments along creeks, improved erosion/runoff control regulation, stormwater infiltration into the ground, and use of additional wetlands throughout the watershed for natural treatment (Irondequoit Basin Project Technical Team 1986).

Pollutant Recycling from Bay Sediments
In June 1981, the Monroe County Health Department and the University of Rochester began a study to determine the rate of nutrient

release from the bottom sediments of Irondequoit Bay. Phosphorus exchange rates between the sediments and the water column were measured, and the results of these flux measurements indicated the majority of nutrients that originate within the bay diffuse from the organic rich sediments of the deeper areas (Burton and Holdren, 1984). This study found that polluted sediments contributed substantial quantities of phosphorus to bay waters and that eutrophication of the bay would continue until the phosphorus-rich sediments were covered by sediments with lower levels of available phosphorus (Irondequoit Basin Technical Group 1986).

Four remedial techniques to resolve the internal recycling problem were evaluated: flocculation and sealing with alum, chemical deactivation of bottom sediment, dredging, and aeration. Flocculation- and sediment-sealing studies using alum alone proved to be the most cost effective method for restoration (Burton and Holdren, 1984). The application of aluminum sulfate to the deeper areas of the bay would precipitate dissolved phosphorus to the bottom, forming an alum blanket which would reduce sediment release of phosphorus. In 1982, alum treatment of the bay was demonstrated in a small area (Ides Cove) (Irondequoit Basin Technical Group 1987).

Agricultural Runoff

A 1984 study by NYSDEC examined the amounts of phosphorus and sediment released from active agricultural land in the rural Thornell subbasin of the watershed, and the effectiveness of alternative agricultural practices in reducing phosphorus from agricultural runoff (NYSDEC 1984). This study concluded that agricultural runoff contributes significant amounts of phosphorus, especially during planting seasons with relatively high rainfall. Several practices were shown to reduce phosphorus runoff (contouring, strip cropping and sodbased rotations) (Irondequoit Basin Project Technical Team 1986).

Water Quality Management Plan

The Water Quality Management Plan Project, led by a Monroe County interdepartmental technical team, integrated the findings of the three major research studies described above into an action plan (Irondequoit Basin Technical Group 1987). A comprehensive "Irondequoit Basin Framework Plan" was prepared, detailing programs and activities of an overall management system for the basin with the county taking a lead role. A companion document, the Irondequoit Basin Policy Report, summarizes the framework plan and outlines a phased implementation strategy to improve water quality in the Irondequoit basin. The county's advisory committee, the Monroe County Water Quality Management Committee, adopted a resolution in February 1986 in support of implementation of recommendations of this policy report (Irondequoit Basin Technical Group 1987).

Irondequoit Basin Plan Implementation Projects

Implementation of the water quality management plan, as outlined in the policy report, began in 1986. The major accomplishments to date are outlined below.

Alum Project
The deep portions of Irondequoit Bay were treated in 1986 with aluminum sulfate at a cost of $660,000 shared by U.S. EPA, State of New York and Monroe County (Irondequoit Basin Technical Group 1987). After the alum treatment, chlorophyll and total algal volume were lowered and hypolimnetic (deep water) phosphorus concentration declined by about 75 percent. Monitoring results through 1988 indicate that these improvements have been maintained through the first two years following treatment (Irondequoit Basin Technical Group 1990).

Irondequoit Bay Drainage Basin Stream Monitoring Program
In cooperation with U.S. GS, Monroe County monitored water quality at six stream sites in the Irondequoit Bay Basin. While baseline phosphorus loadings are generally well below the water quality management plan goal of 14 kg phosphorus/day, phosphorus loadings during storms significantly increase the amount of phosphorus carried to the bay. Now that the alum project has been completed, phosphorus from stormwater contributes 80 to 90 percent of the annual phosphorus load to the bay (Irondequoit Basin Technical Groups 1990).

Atmospheric Deposition Monitoring
Because of the demonstrated importance of air deposition to urban stormwater runoff and because of the problems associated with acid rainfall, atmospheric deposition of selected pollutants has been monitored in the basin since 1980. On an annual basis, atmospheric deposition accounts for 66 percent of the total phosphorus in the stream load. The bulk of the pollutants (about 70 percent of the total phosphorus) falls during dry periods (Irondequoit Basin Technical Group 1990).

Erosion Control Technician Program
The Monroe County Department of Health contracted with the Monroe County Soil and Water Conservation District to provide an erosion control technician to assist in the protection of basin water quality near construction sites. From summer 1988 through the spring 1990, the technician reviewed plans and examined development project sites in the Irondequoit basin. The technician transmitted recommendations for specific sites through the Health Department to municipalities and developers. While municipalities were very cooperative with this effort, the program was discontinued in spring 1990 because of severe county budget constraints. It was reinstituted with state funding to ensure the program continued at least through March 1991.

Continuing Research
The following water quality research studies have been conducted in the basin during the last several years: in-stream impoundments; detention basin stormwater renovation; aquifer research; agricultural land surveys; road salt use survey; salt impacts on Irondequoit Bay; and wetland flow stabilization research.

The detention basin monitoring project is a cooperative effort of the county and the U.S. GS to determine the effectiveness of detention basins to improve stormwater runoff quality under various operating conditions. Stormwater quantity and quality are monitored to determine possible pond modifications to increase the water residence time under varying conditions.

The wetland flow stabilization research has included data collection to understand the function of a wetland south of Irondequoit Bay in detaining and retaining soluable and colloidal phosphorus through wetland biofilm sorption. Recent funding commitments on the part of U.S. GS and Monroe County will provide resources to install a flow control structure to keep storm flows in the wetland longer in order to evaluate enhanced biofilm sorption.

Intergovernmental Agreement

Monroe County has agreed to work together with the town of Pittsford to improve water quality. Specifically, the county and town will cooperatively monitor water quality and quantity in the east branch of Allens Creek (which eventually flows into Irondequoit Creek) and develop a detailed plan for this subwatershed, which will result in improved drainage and water quality. Second, town ordinances that have a positive impact on water quality will be evaluated for opportunities to further enhance water quality. To achieve these goals, 17 specific activities will be undertaken by the town and/or county. These activities include installation and operation of a water quality monitoring station, meetings between the town and the county, and special review of development proposals in the east branch of Allens Creek watershed to identify drainage designs to protect water quality. The town and county also will coordinate responses to water quality complaints in an effort to educate residents about ways they can help improve water quality.

Irondequoit Bay Coordinating Committee

Irondequoit Bay lies below steep wooded slopes, the shoreline is jeweled with hidden coves, and extensive wetlands exist along many areas. Since water quality has improved and its resource value increased, the bay has been under increasing development pressure for homes, recreation, and water-dependent commercial uses. The 1986 opening of a channel from the bay to Lake Ontario heightened the desirability of the bay for development.

The Irondequoit Bay Coordinating Committee was created in 1984 to address the potential problems associated with renewed interest in the sensitive bay area. The committee was created under a cooperative agreement between the County of Monroe, the three major surrounding towns, and the NYSDEC to establish uniform regulations for area development that protect the uniquely sensitive environmental features and thereby decrease the likelihood of further degradation of bay water quality. The committee's work has involved a four-step process: establish environmental objectives; identify appropriate development standards to achieve the objectives; design ordinances and regulations to implement development standards; and recommend a long-range mechanism for continued intergovernmental coordination in the bay area.

By 1985, all four tasks were completed, and by 1990 all the towns had modified their ordinances and regulations to incorporate the majority of the committee's recommendations. Intergovernmental coordination continues through monthly meetings to review all permit applications that are made to the towns, the county, or the NYSDEC.

Waste Site Activities

An interagency committee known as the Waste Site Advisory Committee (WSAC) was formed in 1978 by Monroe County to determine whether or not active or inactive waste sites posed potential environmental or health threats. The WSAC identified approximately 400 waste sites in the county through a survey using aerial photos and agency records. Detailed inventories for seven of the county's 20 towns have been published, and Monroe County and the NYSDEC have written a letter of agreement regarding the New York State Superfund Program, which allows the county to participate in an advisory role (Irondequoit Basin Technical Group 1990).

Remedial Action Plan Preparation

The NYSDEC is responsible for preparing the Rochester Embayment and Lower Genesee River remedial action plan (RAP). In 1987, a consulting firm was hired by U.S. EPA to compile all relevant data and information on the Area of Concern and in early 1989, NYSDEC invited Monroe County to develop a proposal to prepare the Rochester Embayment RAP.

The county believed that watershed plans were needed for each of the three major watersheds that flow to the Rochester Embayment of Lake Ontario. The county proposal was patterned after the Irondequoit basin effort, and proposed a RAP that would consist of four plans—one for each of the three watersheds flowing into the embayment, and one for the embayment itself. A local consulting team is writing the four plans, and the writing is being overseen by an interagency technical group representing county, state, federal agencies and others.

Advisory groups with nearly 200 members have been reorganized to correspond to each of the four plans. The primary advisory group for the Rochester Embayment RAP—the Monroe County Water Quality Management Advisory Committee (WQMAC)—has requested a great deal of background information to help them in the identification of impaired uses. Two community workshops, one on the impact of toxic substances and another on the experiences of other RAPs, have been held to meet the committee's information needs. Both workshops have attracted a great deal of community attention and have been invaluable in focusing RAP efforts.

The WQMAC has gathered information to help determine the extent of use impairments in the Rochester Embayment of Lake Ontario, and the lower segment of the Genesee River that flows into the embayment.

Impairments found in the lower river and the lake include:

- restrictions on fish and wildlife consumption
- degradation of fish and wildlife populations
- degradation of aesthetics and
- loss of fish and wildlife habitat.

For the lake area, there is also evidence of

- some fish tumors
- degradation of benthos
- undesirable algae
- beach closings and
- some taste and odor problems in the drinking water as a result of algae.

Due in part to concerns voiced by the advisory group to the NYSDEC, the state will conduct a special research study on the fish populations in the lower Genesee River. There is some speculation that certain areas of the lower river are devoid of fishes, and a 1991 study will confirm whether or not this is the case and complete further testing if this speculation is proven true.

The three watershed subcommittees of the WQMAC are conducting stream surveys. Since new water quality monitoring was beyond the scope of the original project, subcommittee members decided they should conduct preliminary walks of the streams to identify obvious problems. A technical team set up procedures, a checklist and training programs to conduct the stream survey effort for two of the basins during summer 1990. The results of the summer stream survey work are being summarized, and public participation staff from the Monroe County Department of Planning, as well as technical group members, are assisting with the program.

Public Participation and Intergovernmental Cooperation

Residents in Monroe County have been interested and involved in community issues for a long time, particularly through and including the Rochester Committee for Scientific Information, Sierra Club, the Monroe County Environmental Management Council, university professors, the Monroe County Conservation Council, the League of Women Voters, and many neighborhood associations. In 1980, in response to public interest and public participation requirements for federally funded water quality projects, Monroe County reorganized an earlier advisory group formed as part of the U.S. Clean Water Act planning program (Section 208 of the Clean Water Act) to the Monroe County Water Quality Management Committee (WQMC). Much of the committee's specialized project work was assigned to subcommittees, and these project-related subcommittees advised on the development of the industrial pretreatment program, improvements made to the Van Lare Wastewater Treatment Plant, the design and construction of the combined sewer overflow abatement program, nonpoint source research projects in the Irondequoit basin, and sewer planning.

With initiation of the Rochester Embayment RAP project, the advisory committee was reorganized in 1989 with broader community representation. Twenty-seven members are drawn from four broad categories: citizens, government officials, and economic and public interests, and the committee name has been changed to the

Water Quality Management Advisory Committee (WQMAC). The new structure includes three standing subcommittees, one for each of the three watersheds that flow through the county to the Rochester Embayment of Lake Ontario, and two ad hoc subcommittees to work on the ongoing CSOAP project and Van Lare treatment facilities improvements. Monroe County recognizes that effective public participation will be essential and it will continue to involve citizens in the identification of goals, in the analysis of remedial options, and in the design and construction of remedial actions.

Formal and informal intergovernmental cooperation has also proven essential in evaluating the feasibility of a countywide sewer system. It will be even more crucial in future water quality management activities that may focus on nonpoint source pollution. Much of the nonpoint source pollution is related to development and drainage, where cities, towns and villages have primary authority. Local officials served on the Irondequoit basin subcommittee of the WQMAC, which advised on planning and implementing nonpoint source actions in that watershed prior to the 1989 reorganization, and worked on policy and technical issues.

One challenge is to continue the active involvement of elected officials in resolving technical issues, and thus the advisory group reorganization in 1989 included the creation of a government policy group for elected officials to meet periodically on policy issues related to the RAP. The chief elected official from every city, town and village in Monroe County and every county in the three tributary watersheds serves on this group, where they are advised on major milestones of the project. Monroe County believes that the involvement of elected officials and technical staff from municipalities is essential to the future management of nonpoint sources of pollution, and municipal staff members are becoming involved in all advisory groups and in the RAP project technical group.

Outstanding Technical Obstacles and Challenges

Combined Sewer Overflow Abatement Program (CSOAP)

With the completion of two legs of the CSOAP tunnels, the overflow of combined sewers in the City of Rochester will be reduced from 30 overflow points to only five. The number of times that untreated wastewater will enter the Genesee River as a result of construction is estimated at two times per year, as opposed to the earlier rate of 60 per year. Enlarging sections of the originally proposed tunnel system to improve construction practicality has achieved the desired storage capacity without construction of the three remaining segments. Monroe County is evaluating alternatives to complete the system.

The majority of the completed tunnel system received federal and state aid that reduced the local district cost by 87.5 percent. Available funding to continue this program has been significantly reduced because the grant program has been revised to a revolving loan program that results in local cost reductions of only 15–20 percent. If additional construction is recommended, it will likely result in an increased cost to local district users and thus it will be important to communicate the needs and benefits of such programs to the district users, and to involve the citizens in the decisionmaking process.

Treatment of Combined Sewage

As a result of the CSOAP tunnel program, combined sewage is stored in deep tunnels and later conveyed to the Van Lare Wastewater Treatment Plant. To date, up to 50 percent of the combined sewage routed to the plant receives full treatment, and additional treatment facilities recently put into operation provide primary treatment (screening and grit removal) and chlorination before being discharged through the plant outfall structure. A future challenge will be to optimize pollutant removal.

Sewage Sludge Management

In Monroe County, all sewage sludge from municipal wastewater is burned. Plans are underway to rehabilitate county incinerators to allow this procedure to continue until revised incinerator standards are finalized by U.S. EPA. Identifying an ideal alternative sludge disposal method that protects water and air quality will be a substantial challenge.

Treated Effluent and Industrial Sewage Discharge to
Inland Waterways

Some municipal wastewater and industrial effluent continues to
enter the Genesee River, its tributaries and Northrup Creek. The
industrial pretreatment program is working well in Monroe County,
however continuous and vigorous oversight is required. The challenge will be to continue the development of effective communication between all stakeholders in the community to lead to actions
that will address the interests of the environment and economic
development.

Stormwater Management

An extensive water quality-based management program has been
developed for the Irondequoit Creek Basin to focus on stormwater
management. This plan includes an erosion control technician program, monitoring and research programs, and an intergovernmental
agreement with the town of Pittsford. A primary tenet of this program is to minimize stormwater runoff at the source, using practices such as infiltration to the ground, and the construction of
wetlands that provide natural systems to treat stormwater.

Several challenges remain to implement this program and adequately manage stormwater. For example, changes are required in
the New York State stream classification system and its corresponding stream standards, since they do not currently take into
consideration the significant intermittent water quality problems
associated with storms. More intergovernmental agreements, such
as the one with Pittsford, are needed to foster municipal commitments to stormwater quality management. In addition, innovative
funding mechanisms and a great deal of community support, involvement and communication are needed to procure permanent
stormwater management funding.

Monitoring Efforts

Monroe County conducts extensive water quality monitoring using
funds from the county, state health department, pure waters districts, and U.S. GS. Sustained funding to monitor sewage effluent,

stream flows, groundwater, beaches, and air deposition is always a challenge.

In addition, aquatic and terrestrial indicator species must be identified and their health, abundance and toxic contaminant levels tracked. Behavior and reproductive studies in these indicator species will also be beneficial to measure the success of remedial efforts and identify the need for new actions.

Conclusions

Because water quality in the Rochester Embayment is a reflection of what occurs upstream, the Rochester Embayment RAP consists of four separate planning efforts: one for each of the three subbasins (Lake Ontario West, Central, and Genesee River) and one for the embayment itself. Intergovernmental cooperation and public participation are and have been cornerstones of this work. Monroe County will continually re-evaluate water quality plans and actions based on changes in conditions, technology, community concerns and funding, but a simple process to accomplish this is needed. Future success is dependent on education and funding, and any plans to manage water quality must be cognizant of the overall picture. In order to clean our water, we need to take the ecosystem approach—looking at how people interact with water, animals, plants, earth, and air.

REFERENCES

Burton, R.S. 1975. *A Report on Water Quality at Ontario Beach 1973-1975.* Monroe County Department of Health. Rochester, New York.
Burton, R.S., and R. Holdren. 1984. *Irondequoit Bay–Ides Cove Lake Restoration (314) Diagnostic/Feasibility Study Final Report.* Rochester, New York.
Christensen, N.A. 1968. *Monroe County Comprehensive Sewerage Study, Summary Report.* Rochester, New York.
Davis, J.M. 1989. *Pure Waters Program-Problem/Solution/Results outline.* Rochester, New York.
Day, D. 1967. Transcribed Speech to Rochester Section of the American Chemical Society, *Water Pollution in the Genesee Valley: Past, Present, and Future.* Rochester, New York.
Day, D. 1973. *Letter to Richard F. Scherberger regarding Genesee River Dredging.* Rochester, New York.

Irondequoit Basin Project Technical Team. 1986. *Policy Report, Proposed Approach for Water Quality Management in the Irondequoit Basin.* Rochester, New York.

Irondequoit Basin Technical Group. 1987. *Irondequoit Basin Water Quality Management Annual Progress Report for 1985.* Rochester, New York.

Irondequoit Basin Technical Group. 1990. *Irondequoit Basin Progress Report 1986–1988.* Rochester, New York.

Joint Venture (Erdman Anthony and Knecht Inc.). 1977. *Wastewater Facilities Plan, Combined Sewer Overflow Abatement Program: Special Report.* Rochester, New York.

Kuichling, E. 1907. *Report on the Disposal of the Sewage of the City of Rochester, New York.* Rochester, New York.

Lindsley, J.M. 1974. Secondary Plant: Shoehorned into a small space. *Water and Wastes Engineering.* 11:18–21.

McKelvey, B 1956. The History of Public Health in Rochester, New York. *Rochester History.* Vol. 18, No. 3.

Metcalf and Eddy Engineers. 1929. *Report to Harold W. Baker, Commissioner of Public Works upon Sewage Disposal Problems.* Rochester, New York.

Monroe County Environmental Management Council (EMC). 1983. *Information Report: Genesee River Harbor Maintenance Dredging.* Rochester, New York.

Monroe County Pure Waters Agency. 1969. *Pure Waters Master Plan Report.* County of Monroe. Rochester, New York.

Monroe County Water Quality Management Agency. 1987. *1985–1986 Annual Report of the Monroe County Water Quality Management Agency.* Rochester, New York.

New York State Department of Environmental Conservation (NYSDEC). 1984. *Irondequoit Basin Agricultural Non-Point Source Study.* Technical Report No. 67, Bureau of Water Research, Division of Water. Albany, New York.

New York State Department of Environmental Conservation (NYSDEC). 1990. *Order on Consent, The Eastman Kodak Company.* Rochester, New York.

Quinn, T. 1979. *April 4, 1979 memorandum to Jon Davis.* Monroe County Pure Waters Agency. Rochester, New York.

Rochester Academy of Science. 1982. 1964–1981 Studies of Pollution Control in a Lakefront Community. *Proceedings of the Rochester Academy of Science, Inc.* Rochester, New York.

Rochester Gas and Electric Corporation (RG&E). 1977. *Beebee Station (Station 3) 316(a) Demonstration NPDES Permit No. 070 0X2 2 000103 (NY 0000621).* Rochester, New York.

Rochester Gas and Electric Corporation (RG&E). 1985. *1981, 1982, 1983, 1984 and 1985 Impingement Program, Interim Report: Beebee Power Station.* Rochester, New York.

U.S. Army Corps of Engineers (U.S. ACOE). 1968. *Study of Alternative Disposal Area Rochester Harbor, New York, Preliminary Report.* Buffalo, New York.

U.S. Environmental Protection Agency (U.S. EPA). 1975. *Memorandum of Understanding with U.S. ACOE; Subject: Reconsideration of Diked Area for Disposal of Dredged Spoil from Rochester Harbor.* Buffalo, New York.

Chapter 10

Working toward a Remedial Action Plan for the Grand Calumet River and Indiana Harbor Ship Canal

Michael O. Holowaty, Mark Reshkin,
Michael J. Mikulka, and Robert D. Tolpa

> "The Indiana Harbor Canal and Grand Calumet River is an International Joint Commission Area of Concern and an environmental disaster area. This area represents a challenge of the highest order to the U.S. Environmental Protection Agency, the State of Indiana and their sister agencies. A century of abuse will not be repaired overnight, but all the agencies involved in the RAP have begun the work to restore and protect, for future generations, this vital area of the Great Lakes."
>
> Valdas V. Adamkus
> Region V Administrator
> U.S. Environmental Protection Agency

Introduction

The Grand Calumet River (GCR) and the Indiana Harbor Ship Canal (IHC) comprise a relatively small system in the northwest corner of Indiana (Figure 18). The system is both a unique ecological habitat and an environmental disaster. The river and canal cover only 34 km (21 miles) but have been the focus of attention since 1965 because of the extent of environmental problems. Through the secretary of Health, Education and Welfare (HEW), a public conference on pollution of interstate waters was convened in 1965. The first of several in the area, the conference initiated a coordinated planning, monitoring, and evaluation process to assess the progress of pollution control efforts in southern Lake Michigan, the principle source of potable water for Chicago, Illinois and northwest Indiana (Technical Committee on Water Quality 1970).

A quarter of a century later, the GCR/IHC are still the center of direct attention by the federal government including the U.S. Envi-

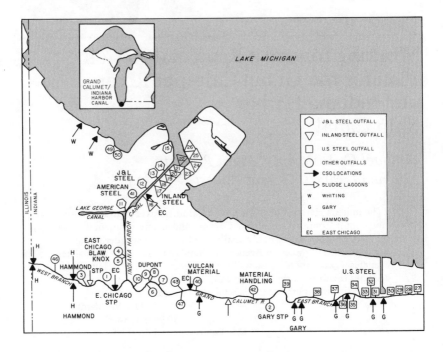

Fig. 18. Industrial and municipal discharges to the Grand Calumet River/Indiana Harbor Ship Canal (U.S. EPA 1985)

ronmental Protection Agency (U.S. EPA) and the U.S. Army Corps of Engineers (Corps). U.S. EPA has been directed under Section 118 of the Clean Water Act, as amended in 1987, to ensure that remedial action plans (RAPs) are developed for all Areas of Concern in the Great Lakes basin, including GCR/IHC. This requirement addresses the United States–Canada commitment to implement the Great Lake Water Quality Agreement, particularly to ensure that RAPs are developed and implemented to restore impaired beneficial uses in Areas of Concern.

The continuing attention to the GCR/IHC stems from the extensive industrialization and modification of the area. It is still home of several major integrated steel mills, a major petrochemical company and three major municipalities. Industrial wastewater comprises over 90 percent of flow for over half of the GCR/IHC's length. Sediments ranging from moderately polluted to toxic reach depths of 5 m (17 feet). The Corps has estimated that in this small river system there may be over 3.06 million m³ (4 million cubic yards) of contaminated sediments (U.S. Army Corps of Engineers 1986).

Although the problems are large in scope, they can be summarized into five general areas.

- Even after all current sources of pollution are abated, additional remediation will be required to restore a balanced ecosystem due to over 100 years of serious environmental degradation.
- A chronic lack of compliance with federal and state environmental statutes by both industries and municipalities.
- The river was literally created by and for heavy industry and compliance with environmental statutes has historically been viewed as costing money, productivity and jobs, by industry as well as area residents.
- The Grand Calumet River has been viewed as a "working river," a euphemism for an open sewer and few—if any—segments of the community envision it as environmentally desirable to restore it for recreational and other uses.
- Northwest Indiana has historically supported the "wrong" political party, from the Governor's view. This has effectively left the area disenfranchised by the state.

In the 25 years since the HEW conference, much has been learned about the river and canal. Numerous studies have been carried out on municipal and industrial discharges, water quality, stream flow, and sediments. Plans have been made by local, state, and federal agencies to address the acute and chronic pollution problems; these efforts continue today and will culminate in the development of a RAP.

The underlying keys to success to RAP development for Northwest Indiana are not technical or scientific. Rather, the successful implementation of a major environmental initiative will be an amalgam of political acumen, the ability to find connections between apparently unrelated environmental problems, and the courage to act despite an adverse or apathetic public reaction.

This chapter will take a broad look at the forces that shaped the area, the role of industry, the byzantine political climate that allowed the current situation to develop, U.S. EPA's enforcement actions, and efforts underway to develop the RAP. The conclusions reflect the personal assessments of the authors and not necessarily the views of U.S. EPA.

Development in the Calumet Region of Northwest Indiana

Natural Setting

The Calumet region is named after the Calumet River and is a low relief area occupying the bed of glacial Lake Chicago. Until the latter part of the 19th Century, the landscape's natural character was barely altered from what existed when the present shoreline of Lake Michigan was formed 2,500 years ago. Three major relict shorelines capped by sand dunes, some rising 12 m (40 feet) above the surrounding plain, represent successively lower stages of glacial Lake Chicago that inundated the plain 14,000 years ago. High and dry, these dunes became major transportation routes through what was an otherwise relatively impenetrable wetland area, first as Native American trails and later as railroads. Settlement of the area surrounding the GCR/IHC resulted in sand mining and removal of all but a few of these shoreline features.

Natural Resource Modification

The first substantial modification of the Calumet region began in 1851 as railroad tracks were pushed through the Calumet Lake plain to link the rapidly growing city of Chicago with older cities such as Fort Wayne, Indianapolis, and eastern seaboard cities. Sand mining and dredging soon became a major activity in the Calumet Lake plain; today, no dune ridges remain in the area surrounding the GCR/IHC.

Drainage was changed most dramatically. More than 150 years ago, Native Americans opened a new channel to Lake Michigan in Illinois by pulling their canoes through the marshland, causing drainage from the west-flowing arm of the Little Calumet River to drain into Lake Michigan about 20 km (12 miles) south of the Chicago River's mouth.

Several drainage changes in Illinois during the 20th Century have shifted drainage from much of the Calumet Lake plain from the Gulf of St. Lawrence to the Mississippi River. The Chicago Sanitary and Ship Canal, completed in 1900, set the pattern for these changes, when it reversed the flow of the Chicago River away from Lake Michigan and southward through the lake plain and morainal area to Lockport, Illinois and the Des Plaines River. The Calumet Sag Canal was completed in 1922, which diverted Little Calumet River flow west of Burns Ditch and the Grand Calumet away from Lake Michigan at Calumet Harbor to the Chicago Sanitary and Ship Canal to the southwest.

At the western edge of the Calumet lacustrine plain, three lakes existed prior to land modification: Wolf, George, and Berry Lakes were remains of a former large bay of Lake Michigan. Only Wolf Lake remains intact today, while Berry Lake was drained to allow for development of Whiting and East Chicago, and Lake George has been filled extensively with slag and sand by adjacent industries.

The construction of harbors, canals, and landfills also began at the dawn of the 20th Century. Prior to this, only the Chicago and Michigan City harbors served the south end of Lake Michigan. Indiana Harbor in East Chicago began with construction of an outer breakwater built some 550 m (1800 feet) out into the lake, while the Indiana Ship Canal was begun in 1903 and was navigable 1.6 km (1 mile) inland by 1909. In 1925, the federal government accepted deed to the completed waterway. The drainage changes of the early 20th Century accelerated and sustained industrialization to the degree that the area today is one of the nation's main urban and industrial regions.

Settlement and Industrialization

The natural environment of the Grand Calumet region changed quickly as the area opened for settlement during the 1830s by European settlers. The fate of the area was decided with the arrival of railroads in the 1850s. Development accelerated, a meat packing plant opened in 1869 and was followed 20 years later by what was to become the world's largest oil refinery, Standard Oil Company. This oil refinery disposed of gasoline high in sulfur content into nearby streams and creeks. This discharge's foul smell outraged the local communities and resulted in the company changing locations, which perhaps represents one of the few environmental community actions in the late 19th Century.

The steel industry entered the Calumet region in 1901 and by 1943, the region supported three large steel companies which included 14 blast furnaces, nine coke batteries and 80 open hearth furnaces. One company, Inland Steel, produced an estimated 181,800 metric tonnes (200,000 tons) of flue dust, 227,300 metric tonnes (250,000 tons) of coal and coke, and 45,500 metric tonnes (50,000 tons) of iron oxide annually, all of which were emitted into the atmosphere. Other waste materials from these facilities included effluents from rolling mills, solid and liquid waste from pickling lines, and slag from furnace operations, which were discharged in wastewater directly to the Grand Calumet River, landfilled or lakefilled. A cement company established in 1903, used blast furnace slag and limestone fines as raw materials to counteract some of the steel industry pollution.

The post 1830s settlement history of the Calumet region can be viewed as a century of growth for many distinct and independent communities, each associated with a particular industry or corporation. This growth pattern led to the balkanized Calumet region of today, an area in which little cooperation exists among communities. Basic infrastructure services such as water supply, wastewater treatment, mass transit, solid waste disposal and shoreline erosion management have become more and more difficult to provide in each community, and thus the optimum time to introduce greatly expanded, substate regionalism for the Calumet area for public works and environmental management may be now.

Federal and State Enforcement

Federal and state enforcement in the GCR/IHC Area of Concern predated the Federal Water Pollution Control Act (FWPCA) Amendments of 1972. This act created the National Pollutant Discharge Elimination System (NPDES) permit program and its commensurate enforcement provisions. Municipal and industrial point sources involved in enforcement actions are provided in Figure 18 (U.S. EPA 1985).

The 1965 HEW conference documented baseline environmental conditions in the Calumet region and established pollution control objectives. In 1970, the Technical Committee on Water Quality for the Calumet area concluded that existing and planned pollution control measures would be inadequate to meet water quality criteria and recommended additional controls (Technical Committee on Water Quality 1970).

Shortly thereafter, the Federal Water Pollution Control Administration was replaced by U.S. EPA, and a nationwide initiative of pollution control was started. As a result, emphasis shifted away from implementation of the technical committee's specific recommendations toward implementation of more generic, broader U.S. EPA water pollution control programs (U.S. EPA 1985).

In retrospect, some of the original recommendations of the technical committee, if pursued, may have resulted in more rapid improvements in water quality. These recommendations included:

1. Initiate a long-range program for recycling and re-use of municipal and industrial wastewater;
2. Eliminate or control combined sewer overflows by July 1, 1977; and
3. Accept industrial discharges only if collection and treatment facilities are available and regulatory approval is obtained.

However, some objectives may never be achieved due to the direction imposed on municipal and industrial waste controls by the NPDES permit program. While the problems facing the area today still generally fall into the three categories presented above, the build-up of polluted sediments from past permitted and unpermitted discharges of municipal and industrial wastes also exists. Federal and state enforcement actions in northwest Indiana have been site-specific and primarily media-specific efforts to comply with NPDES permits, air pollution control requirements and hazardous waste handling requirements, rather than multi-media or multi-facility efforts. This lack of coordination has worked to the detriment of the environment.

Municipal Enforcement

Municipal enforcement in northwest Indiana continues to be a frustrating experience for federal and state enforcement staff and the local citizenry. Actual environmental improvement at the three major municipalities seems to move at a snail's pace. Combined federal and state efforts over the past 25 years are summarized below.

Hammond Sanitary District

The Hammond Sanitary District (HSD) was required to abate polluted storm sewer discharges to Lake Michigan by December 1970, expand the sewage treatment plant, and add advanced waste treatment by December 1972 and, in addition, abate combined sewer overflows to the Grand and Little Calumet Rivers and Wolf Lake by July 1977 (Technical Committee on Water Quality 1970). Failure to complete the storm sewer work by the specified deadline resulted in a 180-day notice by the enforcement division of U.S. EPA in October 1971. This resulted in complete sewer separation of the Robertsdale area of Hammond by May 1973, leaving only a chlorinated storm water discharge to Lake Michigan, and all required sewage treatment plant improvements were completed in 1977 (U.S. EPA 1985).

During summer 1980, Chicago beaches were closed on numerous occasions due to contamination by fecal matter and grease balls. An investigation determined that the source of the material was the Robertsdale Pumping Station of the HSD, and subsequent corrective actions were taken. A second order required the elimination of inflow sources to the sanitary sewer system and the construction of an 8 km (5 mile) forcemain to ensure wastewater from Lever Brothers reached the HSD treatment plant, and included a $50,000 penalty. This work was completed in 1981.

In July 1983, U.S. EPA and the State of Indiana again filed suit against HSD for its discharge of sludge to the Grand Calumet River. As part of a comprehensive settlement entered in District Court in December 1986, the HSD was required to construct and operate adequate sludge handling facilities, close its sludge lagoons, implement its approved pretreatment program, and pay another $50,000 fine. HSD failed to live up to the terms of this agreement and in March 1988, U.S. EPA initiated contempt proceedings against HSD. A contempt hearing was set for November 1991.

Since that time, various measures in secondary treatment and combined sewer overflow controls have significantly improved effluent quality. As a result of a state order, an illegal bypass was eliminated and all flows bypassing the tertiary filters were routed through chlorination and discharged via the main plant outfall. However, numerous additional improvements in pretreatment, the wastewater treatment plant and sewer system are still required and will be specified in a new consent decree.

East Chicago Sanitary District

As a result of the 1965 HEW conference, the East Chicago Sanitary District (ECSD) was required to implement phosphorus controls by December 1972 and advanced waste treatment to eliminate combined sewer overflows and polluted storm water discharges by July 1, 1977 (Technical Committee on Water Quality 1970). The district got off to a good start in 1971 with the construction of a 530,000 m³ (140 million gallon) detention lagoon to hold the diverted flow from several combined sewer overflows for treatment. In addition, phosphorus removal facilities became operational in September 1973, resulting in compliance with the 80 percent removal requirement (Snow 1974).

Economic limitations prevented the ECSD from implementing advanced treatment of its wastes by July 1977. These problems, and a general lack of effective operation and maintenance, resulted in a 1980 civil action by the State of Illinois. A separate civil action was filed by U.S. EPA and the State of Indiana in 1981, and both cases were subsequently consolidated by the court. A preliminary injunction issued in February 1982 required immediate operation and maintenance measures be instituted by ECSD and a subsequent order in November 1982 outlined an extensive operation and maintenance program for East Chicago in addition to constructing advanced waste treatment facilities. A final Consent Judgment required compliance to BOD, suspended solids, phosphorus and ammonia limits by July 1988, and also required ECSD to: implement its approved pretreatment program; submit and implement enforcement response procedures for industrial pretreatment violations; develop a preventative and routine maintenance plan; and pay civil penalties of $160,000 to the United States and court costs of $40,000 to the State of Illinois. There have been several violations of this

Consent Decree. In 1991, agreement was reached with ECSD to pay $30,000 in stipulated penalties for violations of the 1988 Consent Decree.

Since its inception the ECSD has violated certain final effluent limits, which has resulted in stipulated penalties. The ECSD has also petitioned the court for modification of the limits specified in the judgment. Both of these matters are expected to be resolved shortly, and it appears that structural and operational controls implemented to control combined sewer overflows may finally result in loadings that are close to compliance with the terms and conditions of its NPDES permit. Whether these controls are sufficient to meet water quality criteria has yet to be determined.

Gary Sanitary District

Also as a result of the 1965 HEW conference, the Gary Sanitary District (GSD) was required to implement phosphorus reduction by December 1972, and install advanced treatment and eliminate or control combined sewer overflows by July 1977 (Technical Committee on Water Quality 1970). The State of Indiana initiated a civil action against the GSD through its Stream Pollution Control Board (SPCB). Later, the U.S. EPA issued a 180-day notice to GSD to implement phosphorus removal, advanced waste treatment and combined sewer overflow control.

The GSD failed to comply with the July 1977 deadline (Snow 1974), and after a U.S. civil action and a State of Indiana supplemental complaint and motion to intervene in 1978, the GSD agreed to bring its facilities into routine compliance with secondary treatment by April 1979. In June 1983, a final consent decree, specifying the GSD actions required to achieve compliance with advanced waste treatment requirements by August 1983, was entered by the Court to provide long-term compliance. It did not.

The United States filed a motion to enforce the terms of the 1983 judgment in September 1984. Prior to reaching agreement on July 1986, the United States filed a separate civil action alleging that the GSD was improperly storing sludge containing 27,270 kg (60,000 pounds) of PCBs in the Ralston Street Lagoon and needed to implement measures to properly close this sludge lagoon pursuant to the requirements of the Toxic Substances Control Act. A consolidated consent decree was ordered in 1987, which specified that the final limitations of the NPDES permit were to be met immediately and mandated an extensive remedial program covering every aspect of wastewater treatment plant operation and maintenance. Motions dealing with the district's violations of the 1987 decree have been issued in Hammond, Indiana.

As outlined above, the GSD has had the lead role in the 25-year battle for pollution control. However, the federal government is diligently prosecuting its claims, as specified in 1988 and 1989 complaints. The trial started in 1990 with opening remarks, but was suspended while negotiations continued. As of 1991, GSD was in violation of its NPDES permit for lead, chromium, phenolic compounds, ammonia, suspended solids, chlorine, mercury, copper, and biochemical oxygen demand. Agreement in principle has been reached with the city and a final decree sent to GSD for signature in June 1991.

City of Whiting

In response to the 1965 HEW conference, the City of Whiting was required to eliminate combined sewer overflows and polluted storm sewer discharges to Lake Michigan by December 1970 (Technical Committee on Water Quality 1970). Since this deadline was not met, the city was placed on a court ordered schedule to cease its discharges to Lake Michigan by November 1974. A retention basin was constructed, and all flows from Whiting are now diverted to the Hammond Sanitary District (Snow 1974).

Industrial Enforcement

Industrial enforcement in northwest Indiana has a frustrating history as well. The 1965 HEW conference also included requirements for the industrial sector and, while much of the work in response to the 1965 recommendations was completed on schedule, subsequent evaluation showed that industrial improvements were insufficient to meet water quality objectives. Table 15 contains a summary of enforcement efforts over the past 25 years. While enforcement efforts seemingly have been tremendous and resulting water quality has improved, progress has been slow. Remaining con-

TABLE 15. Chronology of Industrial Enforcement Actions in the Grand Calumet Region

Company	Enforcement	Industry Reaction
American Maize	1988—U.S. EPA enforces sewer use ordinance because of substantial organic loads to Hammond Sanitary District	Tighter control of discharge with daily monitoring
American Oil Company	Continues to be suspected source of oils to groundwater of northwest Indiana	Operates system of oil recovery from wells (admits to 63,600 m³ or 16.8 million gallons of petroleum under a 688 hectare or 1,700 acre site)
EI DuPont	1972—consent decree for effluent violation 1991—initiated information request regarding groundwater seeps	1974—initiated compliance program with decreased number of outfalls; no significant violations since 1991—continued cooperation; shares information and collects data
Energy Cooperative	1980—U.S. EPA files suit for violation of NPDES permit 1983—consent decree entered; hazard evaluation performed because of exposed asbestos, improper storage of PCBs, spillage and other improper hazardous waste storage methods 1991—site now owned by the City of East Chicago; City voluntarily negotiated an Agreed Order with Atlantic Richfield Corporation and the Indiana Dept. of Environmental Management for site cleanup	1984—facility abandoned and bankruptcy declared

Inland Steel	1974—State of Illinois files claim for pollution to Lake Michigan water 1986—U.S. EPA files civil action for improper sampling and analysis procedures 1988—consent order requires correct monitoring procedures, compliance with NPDES permit and civil penalty of $100,000 1990—U.S. EPA files civil action for Clean Air Act, Clean Water Act, Safe Drinking Water Act, and Resource Conservation and Recovery Act violations; NPDES violations for ammonia, phenolic compounds, lead and zinc	1974—corrective action taken for coke plants, blast furnaces and pickle liquor waste 1988—implement interim wastewater treatment of carbon filtration and water reduction program for phenol and ammonia violations 1988—violations of the Clean Water Act continue 1991—negotiations currently underway to resolve violations under each federal statute
LTV Steel	1988—U.S. EPA issues administrative action for oil spills and illegal discharges to Lake Michigan 1988—Indiana issues administrative action for effluent violations of lead, zinc and total suspended solids 1988—Indiana files civil action for effluent violations	Corrective action taken for most violations; however, matter not completely resolved; litigation underway LTV has not complied with terms of the order; violations continue

Sources: Quarterly Noncompliance Report 1990; Snow 1974; Technical Committee on Water Quality 1970; personal communication with Howard Chinn, 1990; personal communication with Matthew Scherschel, Indiana Attorney General, 1990.

(*continued on next page*)

(Table 15—continued)

Company	Enforcement	Industry Reaction
United States Steel Corporation	1970 and 1972—Metropolitan Sanitary District of Greater Chicago and State of Illinois petition for injunctive relief because of pollution to Lake Michigan waters 1973—Indiana files civil action for ammonia and cyanide violations 1977—U.S. EPA files civil action for violation of NPDES permit since 1974 1977—consent decree entered for pollution control improvements at Gary Works; civil penalties of $2,925,000 to the U.S. and to the State of Indiana; and $250,000 for study of environmental impacts of dissolved solids on Lake Michigan and $500,000 for additional research and development 1980—amended consent decree to extend time to complete blast furnace recycle system 1988—U.S. EPA issues administrative action for violation of pretreatment requirements for ammonia, cyanide and phenols, failure to report spills and the illegal discharging of wastewater through cooling outfalls; U.S. EPA files civil actions for violations 1988—U.S. EPA issues petition for Discretionary List of Violating Facilities which would exclude USSC from receiving government contracts, grants or loans for at least one year 1990—consent decree entered specifying a $34.1 million settlement ($25 million in in-plant improvements; $7.5 million in sediment characterization and remediation; $1.6 million in penalties); compliance activities underway	1975—request to have action stayed (request denied) 1976—request for review of NPDES permit limitations and conditions with State of Indiana 1979–1985—pollution control measures taken; compliance activities expected to extend into 1995 1990—present—developed and implementing furnace and coke plant management plan; developed and implementing sediment characterization plan; paid civil penalty of $1.6 million; implementing visible oil, blast furnace and coke plant corrective action plans; implementing corrective actions in steel-making and finishing mill areas 1991—to date, compliance with consent decree is being maintained

USS Lead Refining

1985—U.S. EPA and State of Indiana file civil actions requiring control of contaminated runoff

1991—consent decree lodged

1987—facility shut down and claimed bankruptcy

1991—agreement reached among U.S. EPA, Dept. of Justice and USS Lead; no further operations without valid permits; plans to cover hazardous waste piles under review trols have been or will be specified in NPDES permits or consent decrees. Progress to date has only been possible as a result of enforcement and only with relentless enforcement, can further progress be made.

Maintenance and Compliance Dredging

The U.S. Army Corp of Engineers (Corps) has maintained a federal navigation project at Indiana Harbor since 1910. Harbor depths are authorized by statute and range from 6.7 m (22 feet) in the turning basin and Indiana Harbor Canal, to 8.8 m (29 feet) in the outer harbor for navigation purposes.

To maintain authorized depths, periodic dredging is necessary to eliminate shoals in the outer harbor and accumulated sediments in remaining areas. The last such dredging in Indiana Harbor occurred in 1972 and as a result, the navigation channel in the Calumet River branch of the IHC is impassable to commercial traffic. The remainder of the federal channel is passable only by traffic pushing through the soft sediments (U.S. Army Corps of Engineers 1986).

The Corps has not maintained the channel since 1972 due to a lack of a disposal site for the dredged spoils, which have been classified as heavily polluted or toxic according to U.S EPA guidelines. Prior to 1972, the bulk of dredged spoils were disposed of in open water. Much of the Corps' efforts since 1977 has been characterizing the sediments in the project area, researching techniques for dredging and disposal of polluted sediments, and preparing a draft Environmental Impact Statement (EIS) issued in February 1986. The proposed action was to dredge 1 million m^3 (1.3 million cubic yards) of contaminated sediment over 10 years and dispose of it in a confined disposal facility (CDF).

A review of sediment quality data in the Grand Calumet River, the Indiana Harbor Canal and the Indiana Harbor (U.S. EPA 1972, 1977; Kizlauskas 1980; COE 1979, 1983) showed the sediment to be consistently heavily-polluted pursuant to U.S. EPA Region V's 1977 Guidelines for the Pollutant Classification of Great Lakes Harbor Sediments. The EIS recommendation to construct an inlake confined disposal facility (CDF) at Jeorse Park (East Chicago) was not well received by local citizenry and environmental groups, and the Corps' proposed plan quickly became known as "toxic island" to the local public. The Corps' attempts to assuage the public's fears of toxic pollution did not retain any shred of credibility and was not helped by the local citizenry's experience with the Corps' maintenance dredging project and CDF at Michigan City, Indiana. It is doubtful that an inlake CDF for disposal of dredged material from Indiana Harbor can be constructed, and the public has demanded a more intense look at upland sites and consideration of treatment for the dredged material prior to disposal.

Thus, the Corps is re-evaluating its options for disposal of dredged material from Indiana Harbor. Based on its 1988 plan of study, the revised EIS will focus on the alternatives of three upland sites, one inlake site, and taking no action (COE 1988). The Corps has also conducted additional field work on the harbor sediments since 1986, and these data will be factored into its decisionmaking process. Public concerns and those of other governmental agencies will be more fully addressed in the revised draft (personal communication, Jan Miller, COE).

Water Quality Planning in Northwest Indiana

As outlined, water quality planning is not new to the Grand Calumet River. The 1965 HEW conference initiated comprehensive planning to control pollution to the interstate waters of the Grand Calumet River, Little Calumet River, Wolf Lake and Lake Michigan. The passage of the Federal Water Pollution Control Act of 1972 changed water quality planning significantly and the emphasis shifted from a specific geographic focus on the Calumet River system to implementation of national criteria and programs. As a result, cooperation between state pollution control agencies and among neighboring sanitary districts was no longer encouraged as these entities competed for federal monies.

The Northwestern Indiana Regional Planning Commission (NIRPC) was the designated planning agency under Section 208 of the Federal Water Pollution Control Act. NIRPC received state designation in 1975 and produced a plan in 1978 that described baseline conditions for the area's surface water, and looked at population, growth trends and projected municipal water treatment needs. The plan did not address groundwater, contaminated sediments, toxic pollutants in industrial or municipal discharges, or the ultimate cumulative loading to Lake Michigan.

The cooperation that apparently existed between 1965 through 1970 between Indiana, Illinois, the Greater Chicago Metropolitan Sanitary District and the federal government resulted in a bistate commission to formalize the process of addressing the problems in the Calumet area and foster interstate cooperation. This commission, however, died from lack of interest. In addition, the Indiana State Board of Health, the state pollution control agency, removed itself from the Section 208 planning process. This abdication by the state's agency continued until 1986.

Soon thereafter the federal government lost interest and confidence in areawide planning. In the early 1980s, Congress ceased funding of Section 208, and the planning function was then carried out with a portion of the states' allocation of funds for sewage treatment plant construction.

In northwest Indiana the impetus to address the problems affecting the Calumet River system regionally lost momentum. However, as support for areawide planning decreased, environmental activism grew. In 1983, at an International Joint Commission meeting in Indiana, local environmental groups met with U.S. EPA representatives to press their case for cleanup of the Grand Calumet River system. The Administrator for U.S. EPA Region V committed to development of a master plan for the Grand Calumet Area of Concern; this commitment was the impetus for several federal and state reports, significant federal enforcement action, and the template for the remedial action plan process now underway.

The main purpose of the master plan was to implement solutions to long-standing problems (U.S. EPA 1985). It did not carry out extensive data gathering; rather, it looked at the entire Grand Calumet system and sought to resolve basinwide problems. It recognized that the problems could not be solved only by the infusion of federal funds for sewage treatment plant construction, but would require an array of carefully-orchestrated federal and state water pollution control programs (U.S. EPA 1985).

The plan was presented to local communities and affected publics in 1984 and, after inclusion of comments and revision, issued in final form in 1985. U.S EPA Region V tracked the master plan through 1987. Although every item in the master plan's implementation schedule has not been completed, it is still used as a benchmark by U.S. EPA and the state. However, a major flaw of the master plan is that it addresses only the water media and does not factor in air deposition, solid waste programs, or hazardous waste disposal.

In 1986, the State of Indiana transferred pollution control responsibilities from its State Board of Health to a new Indiana Department of Environmental Management (Indiana DEM). The new commissioner of Indiana DEM committed to correcting the master plan's weaknesses and building off the master plan, and developed the Northwest Indiana Environmental Action Plan (EAP). The intent of the EAP was to incorporate air quality, solid waste, and hazardous waste issues (IDEM 1987).

The EAP was finalized in 1987, which provided a snapshot of environmental conditions in the area at the time from the perspective of an aggressive regulatory agency, and delineated what regulatory actions were underway or pending for each medium. The state also attempted to actively involve the public through the establishment of the Citizens' Advisory Committee (CAC) composed of citizens, business leaders, academicians, industrialists, and environmentalists.

Like the 1970 technical report, the Section 208 plan and the master plan, the EAP was flawed. The EAP was not fully implemented, it was not supported at the local level, and it was not consistent with the International Joint Commission's systematic and comprehensive ecosystem approach called for in remedial action plans being developed for all Areas of Concern in the Great Lakes basin.

State staff, in consultation with the U.S. EPA and the International Joint Commission, subsequently developed a draft of a remedial action plan (RAP) that built extensively on the master plan and the EAP. The RAP reformatted and reinforced many of the recommendations of the EAP and the draft RAP was released in March 1988.

The CAC, Indiana DEM, and U.S. EPA met regularly throughout 1988 to polish the draft RAP and focus on outstanding issues, such as dredging and NPDES permit noncompliance. These actions came to an end, however, in November 1988 when the Governor's office changed parties for the first time in 20 years. The Republican-appointed Indiana DEM commissioner left and the CAC limped on for eight more months before dissolving.

Indiana renewed its efforts to complete the RAP in early 1990. A Citizens Advisory for the Remediation of the Environment (CARE) Committee was established to represent the interests and key organizations in the development and implementation of the Grand Calumet River/Indiana Harbor Canal RAP. Indiana DEM appointed a RAP coordinator in mid-1990 and the state has, for the first time in its history, established a regional office outside Indianapolis. A Stage 1 RAP (problem definition and description of causes) was completed in early 1991.

Conclusions

Based on prior experience, the prospects are not good that a RAP for the Grand Calumet/Indiana Harbor Ship Canal will ever be fully implemented. Although most—if not all—of the problems are technologically or scientifically solvable, the political realities of northwest Indiana may very well preclude their resolution. Ultimately, however, U.S. EPA hopes to achieve five broad goals for this Area of Concern.

1. A Stage 2 RAP is completed by the state and submitted to the International Joint Commission.
2. All permittees are in compliance with federal and state statutes or on a court enforceable schedule.
3. Remediation of past environmental degradation is underway or completed within the Grand Calumet River and the Indiana Harbor Ship Canal, including Corps maintenance dredging.
4. The State of Indiana successfully carries out its responsibilities for pollution control and works closely with dischargers to help them remain in compliance.
5. All affected parties are convinced that a viable ecosystem can be reestablished within the Grand Calumet River Basin.

Comprehensive environmental planning is a sound approach if it has the support of the general public, the local industries, municipalities, and local elected officials. These "stakeholders" must, at a minimum, provide tacit support for the process if it is to work. From the perspective of federal regulators, it appears that the process of comprehensive planning for environmental results has not been fully accepted within northwest Indiana; the abysmal compliance records of the area's municipalities and industries attest to this conclusion.

In order to achieve the above mentioned goals and move towards comprehensive environmental planning, several issues must be considered.

1. Gary Sanitary District: With its long history of noncompliance and mismanagement, this district should be dissolved. A new regional sanitary district should be formed that includes the collar communities contracting with Gary for sewage treatment. The new district would own and operate the treatment works, major intercepting sewers, regional pumping stations and forcemains, and aggressively implement a pretreatment program. Local governments would continue to maintain their local sewer system under rules established by the new district. Implementation of this proposal would, in all likelihood, reduce user charges, increase efficiency, reduce the influence of politics in selection of the sanitary district's management staff and increase permit compliance.

2. Hammond and East Chicago Sanitary Districts: The Cities of Hammond and East Chicago share a common boundary and both discharge to the Grand Calumet River. Their sewage treatment plants are less than 3.2 km (2 miles) apart. These two plants, and the contract communities they serve, should be merged into a single sanitary district to increase inline storage capacity and the economies of scale, and improve treatment of combined sewer overflows. In addition, the facilities would obviate the need for construction of additional laboratory facilities and would facilitate implementation of a pretreatment program.

3. The Steel Industry: The state's ability to improve environmental compliance by the steel industries in northwest Indiana is strongly influenced by the political climate of the state. Since the steel mills in the area produce over 25 percent of the United States' raw steel, their importance is recognized. How-

ever, what must change is the steel industry's view that it should be exempted from environmental regulation and control. The RAP must emphasize that the industries must, at a minimum, comply with established environmental standards and also be encouraged to implement recycling where feasible.

4. The Environmental Groups: The environmental groups have been active in convincing Indiana legislators to pass new State Water Quality Standards in 1990 and in focusing attention on the Corps' proposal to dispose of dredge spoils in Lake Michigan. Because the environmental groups often focus on single issues, they do not appear to have galvanized the public into action or to become a major voice in local political decisions. Despite this, the groups must be included in the RAP development process to balance concerns of industrial and economic development interests.

5. Dredging of the Indiana Harbor and Canal: In 1986, virtually every federal, state and local administrative and environmental group turned their backs on the Corps' proposal to dredge the Indiana Harbor and Canal. The plan was criticized as environmentally unsound and insensitive to concerns of local residents. However, contaminated sediments migrating into southern Lake Michigan have not been controlled, and dredging in northwest Indiana is a keystone in the remediation program. The State of Indiana must exert its leadership potential to galvanize the environmental groups, sanitary districts, the industries and the general public behind the Corps and other local proposals to implement remediation, through enforcement or voluntary initiatives. State leadership is the greatest stumbling block to restoring impaired beneficial uses in the Area of Concern.

6. The Role of the U.S. EPA and the State: Initial steps to pollution control have not changed since 1965. All discharges must be brought into compliance with baseline water quality requirements for toxic and conventional pollutants, including compliance with NPDES permits, control of combined sewer overflows, reduction in oil and grease, adequate pretreatment by industrial users, and elimination or treatment of toxic contaminants in industrial discharges. To avoid further repetitions of past mistakes, U.S EPA will attempt to coordinate and

consolidate multimedia enforcement actions and will continue to work closely with Indiana DEM to complete a Stage 2 RAP and to strengthen the state's presence in the area so that it, and not the U.S. EPA, is the principal control agency enforcing environmental statutes and encouraging cooperative environmental management and protection in the Grand Calumet River/Indiana Harbor Area of Concern.

REFERENCES

Indiana Department of Environmental Management. 1987. *Northwest Indiana Environmental Action Plan.* Indiana Department of Environmental Management. Indianapolis, Indiana.

Indiana Department of Environmental Management. 1988. *Draft Remedial Action Plan for the Grand Calumet River/Indiana Harbor Canal Area of Concern.* Indiana Department of Environmental Management. Indianapolis, Indiana.

Kizlauskas, A.G. 1980. *Field Report for Sediment Sampling in the Grand Calumet River and Indiana Harbor Canal, Indiana.* Memorandum to files of U.S. Environmental Protection Agency. Chicago, Illinois.

Quarterly Noncompliance Report (March). 1990. *Indiana Department of Environmental Management.* Indianapolis, Indiana.

Snow, R.H. 1974. *Water Pollution Investigation: Calumet Area of Lake Michigan.* U.S. Environmental Protection Agency, Report Number EPA-905/9–74–001-A. IIT Research Institute, Chicago, Illinois.

Technical Committee on Water Quality. 1970. *Water Quality in the Calumet Area.* Conference on Pollution of Lower Lake Michigan, Calumet River, Grand Calumet River, Little Calumet River, and Wolf Lake, Illinois and Indiana. Gary, Indiana.

U.S. Army Corps of Engineers. 1988. *Indiana Harbor Navigation Maintenance—Plan of Study.* U.S. Army Corps of Engineers, Chicago District. Chicago, Illinois.

U.S. Army Corps of Engineers. 1986. *Draft Environmental Impact Statement: Indiana Harbor Confined Disposal Facility and Maintenance Dredging, Lake County, Indiana.* U.S. Army Corps of Engineers Chicago District. Chicago, Illinois.

U.S. Army Corps of Engineers. 1983. *Indiana Harbor and Canal Analysis of Sediment Samples Collected in 1983.* U.S. Army Corps of Engineers, Chicago District. Chicago, Illinois.

U.S. EPA. 1985. *Master Plan for Improving Water Quality in the Grand Calumet River/Indiana Harbor Canal.* U.S. Environmental Protection Agency Report Number EPA-905/7–84–003C, Region V. Chicago, Illinois.

U.S. EPA. 1972. *Grand Calumet River Sediment Measurement, Collection and Analysis.* U.S. Environmental Protection Agency, Office of Enforcement and General Counsel, National Field Investigations Center. Cincinnati, Ohio.

Chapter 11

Remediating Contamination in the Waukegan, Illinois, Area of Concern

Philippe Ross, LouAnn Burnett, and Cameron Davis

> "The Waukegan RAP must comprehensively identify and systemati-
> cally resolve all the environmental problems in the Waukegan Area of
> Concern. The local economy will not grow or expand until these prob-
> lems are under control. Without the Waukegan Harbor Citizen Advisory
> Group, local citizens could not get to first base on some of these issues.
> The Citizen Advisory Group is absolutely essential for concerned citi-
> zens to raise concerns, contribute to solutions and help generate the
> political will to get on with the job."
>
> Bert Atkinson
> Waukegan Charter Boat Association

Introduction

Waukegan Harbor lies on the western shore of Lake Michigan approximately 60 km (37 miles) north of Chicago, Illinois. The harbor
has gained international notoriety for the level and extent of its PCB
contamination. For example, PCB concentrations in the sediments
of Slip No. 3 (Figure 19) have been found to be as high as 400,000
mg/kg (Mason 1980). Equally well known has been the protracted
cleanup litigation with the industry responsible for the release of
the PCBs (i.e. Outboard Marine Corporation); almost 13 years of
legal battles elapsed before a Consent Decree was entered for PCB
remediation. The purpose of this chapter is to present the history
of the Waukegan Harbor PCB problem, the nature and adequacy of
the remediation plan called for in a 1988 Consent Decree, and the
prospects for the remedial action plan (RAP) process called for in
the Great Lakes Water Quality Agreement (Canada and the United
States, 1987) to systematically and comprehensively restore all impaired beneficial uses in Waukegan Harbor.

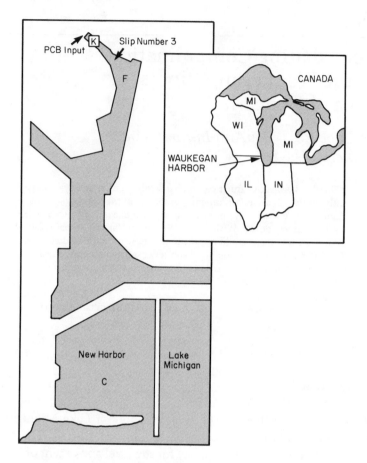

Fig. 19. Waukegan Harbor, Waukegan, Illinois

Background

Geographic Location

The City of Waukegan, Illinois, with a population of about 66,000, surrounds the harbor. The harbor is 1.5 km (0.9 miles) long and runs north and south in an L shape (Figure 19), while two 750 m (2,460 foot) side channels (Slips No. 1 and 3) branch to the west. The harbor has a surface area of 15 hectares (37 acres) and drains an area of 39 hectares (96 acres). Depths range from 2 m (6.6 feet) in the side channels to 8 m (26.2 feet) in the main navigation channel (Thomann and Kontaxis, 1981). The harbor bottom consists of three distinct layers: 1) one to three m (three to 10 feet) top layer of soft sediment/organic silt with a 40–50 percent moisture content; 2) a sand layer; and 3) the natural clay bottom (Mason 1980). The shores of Waukegan Harbor are lined with commercial and industrial facilities, including two Outboard Marine Corporation (OMC) plants, National Gypsum Company, Port Huron Cement, a commercial marina (Larsen Marine) at the north end and a public marina at the south end. Total discharges, including surface runoff and industrial flow, are estimated to be 250 m³ per day. Approximately 750 m³ per day of harbor water is withdrawn by surrounding industries (Thomann and Kontaxis, 1981).

The North Ditch is a small tributary to Lake Michigan located north of Waukegan Harbor, draining 30 hectares (74 acres) of mostly industrial coverage. The flow and depth of water in the North Ditch are governed by rainfall, groundwater, and the level of Lake Michigan. The adverse slope of the bed causes backwater conditions with even minimal wind action. Crescent Ditch and Oval Lagoon at the northwestern extremity of the OMC property, just east of the Eastern Elgin and Joliet Railroad tracks, separate the harbor and industrial area from the rest of the city.

History of Contamination

In 1948, the Johnson Motors Division of OMC, headquartered in Waukegan, began purchasing Pydraul A-200, a hydraulic fluid manufactured by Monsanto Chemical Company. Pydraul A-200 contained polychlorinated biphenyl (PCB) mixtures Aroclor 1242 (66 percent) and Aroclor 1248 (33 percent) and was used in high-pressure diecast machines at OMC's factory on the shore of Waukegan Harbor. These fluids flowed through floor drains into an oil interceptor system that emptied into the North Ditch. Some PCBs escaped from the interceptor system and were released directly to the harbor. Approximately 3,600 metric tonnes (7.92 million pounds) of Pydraul were purchased between 1948 and 1971, and OMC estimates that 400 metric tonnes (880,000 pounds) may have been discharged to the harbor (or approximately 10 percent).

Since 1929, United States industries manufactured and used

PCBs because of their high dielectric constant, their high chemical and thermal stability, their nonflammability, and their low production cost. These characteristics make PCBs highly desirable, but also enable them to accumulate and persist in the environment. Because PCBs are nonpolar, slightly soluble in water, and have a slight vapor pressure, they can be present in the air, water and solid phases of an ecosystem.

There is no substantial evidence that PCBs are produced in the environment, either from natural sources or from the chemical transformation of other compounds, so it must be assumed that all environmental contamination by PCBs is the result of production and use of PCB-containing materials. In 1971, after evidence accumulated that exposure to PCBs could result in hazards to human health and to the environment, Monsanto voluntarily suspended sales of hydraulic fluids like Pydraul. In 1976, the manufacture, processing and distribution of PCBs were banned in the United States, except as authorized by the United States Environmental Protection Agency (U.S. EPA).

Samples taken by the Illinois Environmental Protection Agency (Illinois EPA) in 1976 showed that high levels of PCBs had accumulated in the sediments of Waukegan Harbor; PCBs have since been detected in Waukegan Harbor by several studies (Table 16). The best estimate of the PCB mass residing in the sediments of Waukegan Harbor and the North Ditch is 484,500 kg (1,065,900 pounds). About 43 percent of this mass is in the harbor, and approximately 95 percent of the harbor mass is in Slip No. 3. Present PCB concentrations in the water column vary from 0.6 μg/l in the inner harbor to less than 0.1 μg/l in Lake Michigan immediately outside the harbor. For the North Ditch, data obtained in 1979 indicate average water column concentrations of about 7 μg/l with peak values during rainfall events of 80 to 160 μg/l. The average PCB concentration in open Lake Michigan is approximately 1.2 ng/l and nearshore values average 3.2 ng/l, while values averaging 7.9 ng/l were measured in Lake Michigan near Waukegan Harbor (Swackhamer and Armstrong, 1987).

Although this measurement is the highest water column reading, this does not necessarily reflect a large PCB loading to Lake Michigan, as exchange between the harbor and the lake is low (Swackhamer and Armstrong, 1988). Thomann and Kontaxis (1981) calculated that the net exchange of PCB from the harbor to the lake is about 10 kg (22 pounds) per year, including transient storm events, and that approximately 5 kg (11 pounds) per year of PCBs are discharged from the North Ditch. The flux from Waukegan Harbor and the North Ditch to Lake Michigan as a whole is less than two percent of the total current PCB load of 1,400–5,600 kg (3,080 to 12,320 pounds) per year (Thomann and Kontaxis, 1981).

The flux from Waukegan Harbor to Lake Michigan is, however, more important on a local, nearshore basis. Swackhamer and Armstrong (1988) reported that the impact of Waukegan Harbor on PCB levels in lake sediments within a 10 km (6.3 mile) radius of the harbor is significant. Values along the Lake Michigan shore north of the harbor are generally higher than the overall local mean, while values south of the harbor mouth are lower. Lacking other significant local sources, this suggests either that PCBs are transported from the harbor mouth in a northerly direction, which is consistent with known lake circulation patterns, or that PCBs are discharged from the North Ditch. Localized "hot spots" of sediment

TABLE 16. Summary of PCB Concentrations in the Water Column and Sediment of Waukegan Harbor

Study	Year	PCB in Water Column (µg/L)	PCB in Sediments (mg/kg)
Soil Testing Service	1976	—	<0.1–1.1
ENCOTEC	1976	—	62.0–9,900
OMC	1976	0.22–0.51	—
ENCOTEC	1977a	—	65.0–8,400
ENCOTEC	1977b	0.62–14.0	—
OMC	1978	0.40–1.7	—
ERG, Inc.	1979	0.015–0.087	—
Armstrong	1980	—	181.5–3,634
Mason	1980	—	10.0–400,000
Ross et al.	1988	—	5.0–17,251

with high levels of PCBs may be naturally occurring deposition zones or locations where material dredged from the harbor was dumped (Swackhamer and Armstrong, 1988).

The role of Waukegan Harbor as a point source of PCBs to Lake Michigan was further suggested by comparing the PCB distribution and Aroclor composition of harbor sediments to those of sediments in nondepositional zones at varying distances from the harbor. The Aroclor composition of sediments in Lake Michigan tends to more closely resemble that of harbor sediments within a 10 km (6.3 mile) radius of the harbor mouth than it does elsewhere (Swackhamer and Armstrong, 1988).

Sediment PCB concentrations (Table 16) have been measured from below 0.1 mg/kg (Soil Testing Service 1976) to 400,000 mg/kg (Mason 1980). It should be noted that while some of the variation inherent in this broad range is due to spatial gradients, differences in sampling protocols (grab samples vs. cores, surface samples vs. deep samples), and in analytical methods, have undoubtedly created variations. There is a marked gradient in these concentrations, decreasing from the head of Slip No. 3 into the main harbor and along the Federal Channel to the open lake (Ross et al. 1988).

All sediments with PCB concentrations greater than 500 mg/kg are found in Slip No. 3 (Figure 20), and the zone with sediments in the 50–500 mg/kg range is considered to lie between Slip No. 3 and Slip No. 1 (Figure 20). A battery of toxicity tests performed on sediment extracts from 24 stations throughout Waukegan Harbor showed at least moderate toxicity at all stations, and high toxicity at many. While PCBs clearly play a role in toxicity of these sediments, high overall toxicity is not restricted to areas with high PCB concentrations (Ross et al. 1988). A multitrophic level evaluation by Burton et al. (1989) indicated that Waukegan Harbor sediments are highly toxic to a variety of test organisms, (*Daphnia magna, Ceriodaphnia dubia, Selenastrum capricornutum, Photobacterium phosphoreum* and *Panagrellus redivivus*).

Remedial Action in Waukegan Harbor

History of Negotiations

In the mid-1970s there was growing concern about the dangers of PCBs. When a series of 1976 reports indicated high PCB levels in

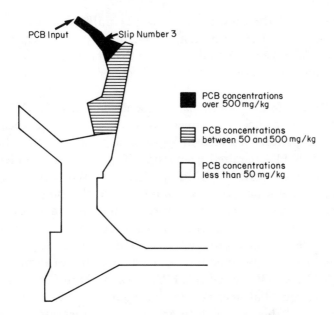

PCB Input — Slip Number 3

■ PCB concentrations
over 500 mg/kg

▤ PCB concentrations
between 50 and 500 mg/kg

☐ PCB concentrations
less than 50 mg/kg

Fig. 20. Map of Waukegan Harbor, Illinois, showing zones of sediment
contamination

and around Waukegan Harbor, state and federal regulatory agencies
began conducting investigations. In 1978, U.S. EPA filed suit against
OMC to clean up PCB contamination in the harbor and on OMC
property, and thus began an adversarial relationship between U.S.
EPA and OMC that would last for eight years. It is unclear what, if
any, remedial activity would have resulted from the 1978 U.S. EPA
lawsuit if there had been no change in the existing environmental
statutes. In 1980, however, U.S. Congress enacted the Comprehensive Environmental Response, Compensation, and Liability Act
(CERCLA, also known as Superfund), enabling the federal government to finance the cleanup of hazardous substances and to pursue
potentially responsible parties for compensation.

In 1982, the OMC/Waukegan Harbor site was included on the
first National Priorities List, which qualified the site for federal
Superfund cleanup funds. A feasibility study examining various remedial options was completed by U.S. EPA and released for public
comment in 1983. In 1984, one alternative was selected by U.S. EPA

in a Record of Decision authorizing $21 million for the cleanup program, and the U.S. Army Corps of Engineers (Corps) began design work for the project. It should be noted that the decision was essentially a unilateral action, as OMC still contested any liability on its part.

In 1985, U.S. EPA's original (1978) lawsuit against OMC was dismissed but the decision allowed for a future suit, under the provisions of Superfund, to recover government costs for the cleanup activity. U.S. EPA required access to OMC property to complete the remedial design work, but OMC denied access. When U.S. EPA was granted an administrative warrant to enter the property, OMC quickly obtained a stay of this warrant and a protracted series of legal proceedings resulted. In 1986, U.S. EPA appealed the denial of access to OMC property to the U.S. Supreme Court. Later that year, Congress enacted the Superfund Amendments and Reauthorization Act (SARA), which granted U.S. EPA access authority to implement Superfund remedial actions. At this point, U.S. EPA and OMC agreed to end their ongoing litigation over site access and began negotiations to cleanup the site. The negotiations between U.S. EPA, the Illinois EPA and OMC continued until a consent decree recommending remedial procedures, was published in 1988. After notice, a public comment period and response, the Consent Decree was finally entered in the U.S. District Court (Northern Illinois) on April 21, 1989.

The 1988 Consent Decree

The remedial action proposed in the 1988 Consent Decree deals with the same areas as the 1984 Record of Decision (Slip No. 3 and the Upper Harbor, the North Ditch; the Crescent Ditch/Oval Lagoon area on OMC property; and the OMC parking lot), and these areas are referred to here as the operable unit. Briefly, eight sequential steps will be taken to accomplish remedial objectives.

1. A slip will be built on the east side of the Upper Harbor to replace Slip No. 3. Larsen Marine's operations will be relocated there (see Figure 21).
2. A double sheet pile cutoff wall will be built to isolate Slip No. 3 from the Upper Harbor. After isolation is completed, a

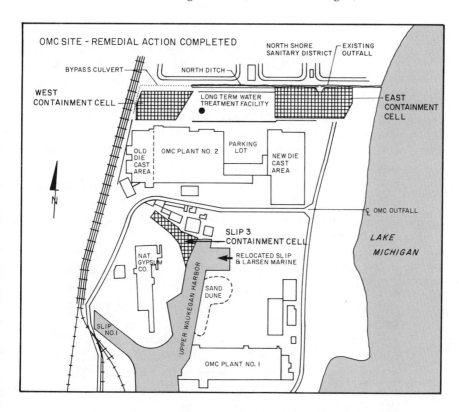

Fig. 21. Projected changes in Waukegan Harbor, Illinois, after completion of Remedial Action proposed by the 1988 Superfund consent decree

water-tight clay slurry wall with a minimum thickness of one meter (3.3 feet) will be anchored to the underlying clay till and a permanent containment cell will be constructed within the slip.

3. Sediments in Slip No. 3 with PCB concentrations above 500 mg/kg will be removed and treated. Sediments in the Upper Harbor containing material with PCB concentrations between 50 and 500 mg/kg will be removed and placed in the new Slip No. 3 containment cell.

4. Using the same design as the Slip No. 3 containment cell, two other cells will be built. The East Containment Cell will in-

clude part of the OMC parking lot and some land east of the lot, while the West Containment Cell will encompass the Crescent Ditch and Oval Lagoon area. Before construction of these cells, all soils with PCB concentrations over 10,000 mg/ kg will be removed for treatment.

5. Material to be removed from designated "hot spot" areas (Slip No. 3, North Ditch, Crescent Ditch and Oval Lagoon) will be treated by the Taciuk process, an anaerobic thermal extraction procedure developed in Alberta, Canada for the commercial removal of crude oil from naturally occurring oil-bearing sands. Since the PCBs at Waukegan tend to be contained in an oily matrix, the Taciuk process is appropriate to extract PCB oil from soils and sediments. At least 97 percent (by mass) of the PCBs will be removed by this process.

6. Extracted PCB oil will be removed offsite for incineration or disposal in U.S. EPA approved facilities. The treated sediments and soils will be placed in the upland containment cells.

7. When all relocated and treated materials have been deposited in containment cells, the cells will be closed and capped with a high-density polyethylene liner and a soil cover. The cells will include extraction well systems to prevent migration of PCBs from the cells.

8. Wastewater generated during construction will be processed onsite in a temporary water treatment facility. Dredging water and wastewater from remedial processes will be treated by sand filtration to remove sediment particles, followed by carbon absorption to remove contaminants. A small, permanent water treatment facility will be built to treat water from the extraction wells in the containment cells.

It should be noted that, at the time of this writing, the actual cleanup is stalled at the first step. The excavation of soil necessary to construct the new slip has been blocked by the discovery of high levels of contamination in these soils. The groups responsible for carrying out the Consent Decree have thus been forced to redesign the disposal plan for these soils before the project can proceed. Construction began in 1990.

Progress Since the 1984 Record of Decision

The 1988 Consent Decree represents an improvement over the 1984 Record of Decision because the new design differs in four major aspects. First, the 1988 decree provides for the replacement of Slip No. 3 with a newly constructed slip, where Larsen Marine will relocate its operations. Second, the 1988 decree expands the definition of the "hot spot" in Slip No. 3 to include all sediments with PCB concentrations above 500 mg/kg, thereby including a much larger amount of material to be treated. Third, containment cells are to be constructed differently (synthetic and soil caps, extraction wells) and will be built inground rather than above ground. Finally, the "hot spot" material will be treated onsite rather than transported offsite to a licensed PCB landfill. This eliminates the need for the dewatering lagoons called for in the 1984 design.

Scope of Current Remedial Actions

The 1988 Consent Decree was a Superfund action, driven primarily by CERCLA and SARA legislation. The proposed remedial action is thus limited, both in area and completeness, to enforceable provisions of those acts. Waukegan Harbor is, however, subject to several other agreements and statutes.

The Great Lakes Water Quality Agreement
Under the auspices of the Great Lakes Water Quality Agreement, Waukegan Harbor is one of 43 Areas of Concern where beneficial uses are impaired. Annex 2 in the Agreement defines the criteria necessary for the removal of a waterway from the list of Areas of Concern, and stipulates that the Parties (i.e. Canada and the United States) shall assist state, provincial and local governments in developing and implementing remedial action plans (RAPs) to restore impaired beneficial uses. Annex 2 also defines the components of a RAP and requires that the public be consulted in the RAP development process. In addition, Annex 14 guides the Parties on research and technology programs to remediate contaminated sediments.

The Federal Water Pollution Control Act ("Clean Water Act") of 1987

The act offers guidance on how U.S. EPA and state agencies should proceed to remediate an Area of Concern such as Waukegan Harbor. Section 118 gives the force of United States law to the Great Lakes Water Quality Agreement, mandating that the federal government, and by extension the states, must "seek to attain the goals embodied in" the Agreement. Section 404 regulates operations for dredging and filling in navigable waters of the United States; this section also affects Waukegan because most of the harbor is a federal channel and thus subject to periodic maintenance dredging.

The Toxic Substances Control Act (TSCA)

This act provides guidance to U.S. EPA for the cleanup of contaminants, including PCBs. TSCA determines the extent of cleanup and the methods to be employed.

Adequacy of the Consent Decree Cleanup Plan

The Superfund cleanup plan as embodied in the 1988 Consent Decree is commendable in that its cost will be funded in large part by the responsible party (OMC), and the cleanup plan will employ treatment technologies rather than disposal measures for most contaminated sediment. It must, however, be viewed as a starting point, not as the last word on remedial action in Waukegan. While the plan may satisfy the requirements of CERCLA, it clearly falls short of objectives contained in other laws and agreements. These deficiencies can be found in three primary areas.

1. Removal of PCBs in the Operable Unit

Although the 1988 Consent Decree cleanup plan will remove greater than 96 percent of the total mass of PCBs in the entire harbor area, the remaining four percent could represent concentrations high enough to present an ecological hazard. The driving principle behind the plan is a PCB target level of 50 mg/kg, which does not take into account the bioaccumulative nature of PCBs. While the cleanup operation is likely to also remove some material with PCB concentrations below 50 mg/kg, it could be problematical for U.S. EPA to return at a later date and remove material with less than 50 mg/kg PCBs at the expense of OMC, since U.S. EPA's inability to pursue cleanup below 50 mg/kg is determined by law. This limit places a severe restriction on the agency's ability to adopt ecologically-sound remedies.

In selecting a cleanup target level of 50 mg/kg, U.S. EPA's Superfund office relied heavily on a prediction (Thomann and Kontaxis, 1981) that such action would result in a negligible flux of PCBs to Lake Michigan. In espousing this rationale, however, the Superfund office fails to pursue a goal that another branch of the agency, the Great Lakes National Program Office (GLNPO), is directed to achieve. As an Area of Concern, Waukegan Harbor itself must eventually be remediated to meet the criteria for delisting set by the Great Lakes Water Quality Agreement. Supporting use of the ecosystem approach in solving contamination problems, the Agreement guides the Parties to ensure that the Great Lakes system is "free from materials... that alone, or in combination with other materials, will produce conditions that are toxic or harmful to human, animal, or aquatic life..." (IJC 1987). As the Agreement defines the Great Lakes system as all "bodies of water that are within the drainage basin" of the Great Lakes (IJC 1987), Waukegan Harbor must be considered a part of that system. Thus U.S. EPA, particularly GLNPO, has a mandate to ensure that the harbor itself, and not just Lake Michigan, is free from toxic substances.

2. PCB Contamination Beyond the Operable Unit

The Superfund action restricts the scope of cleanup operations in Waukegan Harbor to Slip No. 3 and the Upper Harbor. Data from U.S. EPA (1981) and more recent sources (Ross et al. 1988) indicate that PCB contamination above 10 mg/kg also exists in sediments elsewhere in the harbor, outside the operable unit covered by the Consent Decree. If remediation of Waukegan is limited to the Superfund cleanup plan, this would be in marked variance with RAPs from other Areas of Concern that are not Superfund sites. For example, the RAP for the Green Bay/Lower Fox River Area of Concern (see Chapter 2) recommends a target PCB level of no more than 1 mg/kg (Wisconsin DNR 1988) and the U.S. Fish and Wildlife Service has urged PCB cleanup in the Kalamazoo River (Michigan) Area of Concern (see Chapter 12) to no more than 0.05 mg/kg to protect aquatic life.

Although the U.S. Food and Drug Administration consumption threshhold for PCBs in fish tissue is 2 mg/kg, the Great Lakes Water Quality Agreement (IJC 1987) recommends a level of 0.1 mg/kg. It is therefore possible that more stringent standards may one day be adopted by the FDA, in which case current remedial design concepts for Waukegan Harbor could become inadequate to protect human health. The Waukegan situation is further complicated by the fact that maintenance dredging activities, under the auspices of the U.S. Army Corps of Engineers, will eventually be necessary in Slip No. 1 and other parts of the harbor; in that event, disposal of dredged material could become a problem. IJC (1982) dredging guidelines consider sediments with PCB concentrations greater than 10 mg/kg to be polluted. If this standard is applied to dredging operations in Waukegan Harbor, further remedial measures will have to be taken. As the 1988 Consent Decree contains a provision exempting OMC from any further liability, the cost of the future measures will be borne by taxpayers.

3. Contaminants Not Addressed by the Superfund Cleanup Plan

The Superfund action is based entirely on PCB concentrations. Recent data (Risatti et al. 1990) indicate that potentially toxic levels of heavy metals, oils and grease, and polynuclear aromatic hydrocarbons (PAHs) occur in harbor sediments, and are not limited to the Superfund operable unit area. Furthermore, bioassay results indicate high levels of toxicity throughout the harbor, which are not correlated with PCB concentrations (Ross et al. 1988; Burton et al. 1990). This is not surprising because PCBs, while linked to cancer, birth defects, and other health hazards of a chronic nature, are not known for their acute toxicity. Thus, it is reasonable to assume that the observed acute toxicity is caused by other compounds and eventually must be addressed.

Conclusion

It is unrealistic to assume that fulfillment of the 1988 Superfund Consent Decree for the operable unit will be sufficient to delist Waukegan Harbor as an Area of Concern. The RAP development process has already begun (i.e. citizen's committee formed in

August 1990), and the RAP will undoubtedly contain recommenda-tions for further remedial action at further cost to the taxpayer. This situation could have been avoided if it had been possible for the Superfund action to incorporate federal responsibilities under other statutes and the Agreement. The failure to do so must, in part, be attributed to the failure of current laws to keep pace with our knowledge of the fate and effects of PCBs and other contaminants.

In environmental remediation actions, public participation is ob-viously crucial to generate a broad base of community support and to maintain pressure on government agencies to develop an effec-tive RAP. Just as important, public participation also helps to en-sure that a wide variety of concerns can be considered. Because the Superfund Consent Decree was part of a legal enforcement action, opportunities for substantive public participation were severely re-stricted. The RAP process offers great hope that the limitations of the Superfund action will eventually be addressed, but it will be successful only if effective leadership and concerted public action continue to develop. It is incumbent upon state and federal agen-cies, as well as public interest groups such as the Lake Michigan Federation, to facilitate public participation in order to develop a RAP that takes into account a broad range of environmental issues.

REFERENCES

Armstrong, D.E. 1980. *Sediment sampling, water sampling and PCB analysis in Lake Michigan.* Final report submitted to JRB Associates, Inc. McLean, Virginia.

Armstrong, D.E. and D.L. Swackhamer. 1983. PCB accumulation in southern Lake Michigan sediments: evaluation from core analysis. *In:* D. Mackay, S. Peterson, S.J. Eisenreich and M.S. Simmons, editors. *Physical Behavior of PCBs in the Great Lakes.* Ann Arbor Science. Ann Arbor, Michigan. pp. 224–229.

Burton, G.A., Jr., B.L. Stemmer, K.L. Winks, P.E. Ross and L.C. Burnett. 1989. A multitrophic level evaluation of sediment toxicity in Waukegan and Indiana Har-bors. *Environmental Toxicology and Chemistry.* 8:1057–1066.

Canada–United States. 1987. *Great Lakes Water Quality Agreement of 1978, as amended by Protocol signed November 18, 1987.* Windsor, Ontario, Canada.

ENCOTEC. 1976. *Sediment analysis for Waukegan Harbor.* Report to Martin, Craig, Chester, and Sonnenschein. Ann Arbor, Michigan.

ENCOTEC. 1977a. *Waukegan Harbor Sediment Survey.* Report to Martin, Craig, Chester, and Sonnenschein. Ann Arbor, Michigan.

ENCOTEC. 1977b. *Water quality surveys, Waukegan Harbor.* Report to Martin, Craig, Chester, and Sonnenschein. Ann Arbor, Michigan.

ERG, Inc. 1979. *Sampling and analysis of water and sediment samples taken from Waukegan Harbor before, during and after maintenance dredging.* Report to the U.S. Environmental Protection Agency, Region V. Chicago, Illinois.

International Joint Commission (IJC). 1982. *Guidelines and Register for the Evaluation of Great Lakes Dredging Projects.* Great Lakes Water Quality Board. Windsor, Ontario, Canada.

Mason, Silas Co., Inc. 1980. *Plan for the removal and disposal of PCB contaminated soils and sediments at Waukegan, Illinois.* Report prepared for the U.S. Environmental Protection Agency, Region V. Chicago, Illinois.

OMC. 1976. *Harbor water PCB concentration.* Johnson Outboards office memorandum. Waukegan, Illinois.

OMC. 1978. *PCB analysis.* Johnson Outboards office memorandum. Waukegan, Illinois.

Risatti, J.B., P.E. Ross and L.C. Burnett. 1990. *Amendment: Assessment of ecotoxicological hazard of Waukegan Harbor sediments.* Illinois Hazardous Waste Research and Information Center. Report No. HWRIC-RR-052. Champaign, Illinois.

Ross, P.E., M.S. Henebry, L.C. Burnett and W. Wang. 1988. *Assessment of the ecotoxicological hazard of sediments in Waukegan Harbor, Illinois.* Illinois Hazardous Waste Research and Information Center Report No. HWRIC-RR-018. Champaign, Illinois.

Soil Testing Services. 1976. *Summary of sediment sampling, Waukegan Harbor.* Report to Martin, Craig, Chester and Sonnenschein. Northbrook, Illinois.

Swackhamer, D.L. and D.E. Armstrong. 1987. The distribution and characterization of PCBs in Lake Michigan water. *J. Great Lakes Res.* 13:24–36.

Swackhamer, D.L. and D.E. Armstrong. 1988. Horizontal and vertical distribution of PCBs in southern Lake Michigan sediments and the effect of Waukegan Harbor as a point source. *J. Great Lakes Res.* 14:277–290.

Thomann, R.V. and M.T. Kontaxis. 1981. *Mathematical modeling estimate of environmental exposure due to PCB contaminated harbor sediments of Waukegan Harbor and North Ditch.* U.S. EPA Contract Report No. 68-01-3259. HydroQual, Inc. Mahwah, New Jersey.

United States Environmental Protection Agency. 1981. *The PCB contamination problem in Waukegan, Illinois.* Waukegan, Illinois.

Wisconsin Department of Natural Resources. 1988. *Green Bay/Lower Fox River Remedial Action Plan.* WR-175-87. Madison, Wisconsin.

Chapter 12

Remediating Polychlorinated Biphenyl Contamination in Kalamazoo River/Portage Creek Water, Sediment, and Biota

Chris Waggoner and William Creal

> "The Kalamazoo River continues to be a major source of PCBs to Lake Michigan. It is essential that the Kalamazoo RAP is developed in a timely fashion, that it systematically identifies and implements all necessary remedial actions, and that citizens are actively involved throughout the process."
>
> Glenda Daniel
> Executive Director
> Lake Michigan Federation

Introduction

The Kalamazoo River is a unique and valuable waterway that drains almost 5,180 km² (2,000 square miles) from 10 counties in southwestern lower Michigan and empties into southeastern Lake Michigan at the City of Saugatuck, Michigan (Figure 22). Recreational resources in the drainage basin are diverse, ranging from small, high quality trout streams to marshy, wide river mouth areas. The main stream of the Kalamazoo River is a cool-water fishery, with all but a few tributaries designated as trout streams. The lower portion of the main stream is so highly valued as a resource that it has been designated a "Natural River" area under Michigan State Law. This lower portion also supports a good cool water fishery with salmon and steelhead, and wildlife from the Allegan State Game Area.

Although the Kalamazoo River has long been known for its fishing, the river suffered severe water quality degradation during the region's industrial development and growth in the first half of this century. These problems were especially severe in the main stream from Battle Creek to Allegan as a result of poor waste disposal practices. In 1956, the combined pollution from domestic and in-

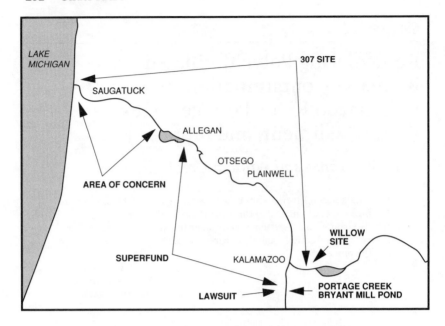

Fig. 22. Areas covered under state (Act 307, lawsuit), federal (Super-fund) and international (International Joint Commission Area of Concern) programs

dustrial sources was equal to the raw sewage waste from one-half million people. Large sections of the river were devoid of the oxygen necessary to support fish, fish kills occurred, and the river gained national prominence when a local fish kill was depicted as "Photo of the Week" in Life Magazine in 1953. Fish populations historically have also been extremely stressed and diseased, including carp with severely eroded fins due to bacterial infections.

In addition, odors from the river forced people living near the river to sleep with their windows closed on hot summer nights. Many residents remember the river turning a milky white color from inadequately treated wastes, and nuisance algal growths were common downstream of the City of Kalamazoo.

The many years of excessive pollution resulted in a massive cleanup effort for the Kalamazoo River in the 1950s, which continues today. Industrial dischargers along the river have either built their own waste treatment plants or have been connected to the municipal wastewater treatment plants. Municipal wastewater treatment plants have vastly improved their treatment systems, and over $500 million (much higher if industry costs are included) have been spent over the past 20 years to handle the domestic and industrial wastes in the Kalamazoo River basin.

As a result, water quality has improved dramatically. The turbid, oily and generally disgusting conditions that persisted through the early 1970s are no longer present, and oxygen levels in the river have improved significantly. Recent benthic macroinvertebrate surveys indicate that water quality is dramatically improved (Heaten 1990, Morse 1990) and the Kalamazoo River now supports healthy populations of game fish such as smallmouth bass, walleye, and brown trout.

The outstanding pollution problem in the Kalamazoo River is PCB-contaminated sediments as a result of historical discharges from paper deinking operations. These inplace contaminated sediments provide an ongoing source of PCBs to the water column, biota, and ultimately Lake Michigan. Michigan Department of Natural Resources (DNR) efforts to coordinate and ensure complementary and reinforcing state (Public Act 307), federal (U.S. Comprehensive Environmental Response, Compensation and Liability Act or Superfund) and international (remedial action plans to restore Great Lakes Areas of Concern) programs to remediate this contamination are outlined in the remaining portion of this chapter.

Remediation of the PCB Problem

Approximately 159,090 kg (350,000 pounds) of PCBs contaminate a 128 km (80 mile) reach of the Kalamazoo River from the City of Kalamazoo to Lake Michigan. The Michigan Department of Public Health advises no human consumption of carp, suckers, catfish, and largemouth bass in the Kalamazoo River (downstream from Morrow Pond Dam to Lake Michigan) and Portage Creek (downstream from Monarch Mill Pond) due to PCB contamination. The lower 41.6 km (26 miles) of the Kalamazoo River are designated an Area of Concern.

PCBs were released into the river through discharges from paper industries and disposal of contaminated material into and adjacent to the river. Effluent from recycling of paper, especially carbonless copy paper during the mid-1950s through early 1970s by local paper companies, contributed the majority of the PCBs, since the chemical was commonly used in carbonless copy paper. The open use of PCBs was virtually banned by the State of Michigan in 1976, and nationwide by the U.S. Toxic Substances Control Act in 1977. The National Pollutant Discharge Elimination System (NPDES) program has provided an additional regulatory tool to severely restrict and virtually eliminate point source discharges of PCBs to the Kalamazoo River.

Because PCBs are insoluble in water and tend to attach to fine sediments like clay and silt, they have concentrated in the sediments and provide a continuing source of PCBs to the water column and food chain. Remedial actions thus are focusing on eliminating the sediments as a source of PCBs to the water column and biota.

Remedial investigations indicate that PCB-contaminated sediments are concentrated in eight primary areas: Bryant Mill Pond on Portage Creek and seven areas on the Kalamazoo River—the Willow Boulevard site, Plainwell Dam, Otsego City Dam, Otsego Dam, Trowbridge Dam, Allegan City Dam and Lake Allegan (Figure 23). Seven areas are sites where dams were used to impound water on the river, thereby slowing the current speed and allowing fine sediments carried in the water to settle out. Interim remedial actions are planned for five areas of contamination, and additional studies are planned for all eight areas prior to final remediation.

The goal of remedial actions is to meet Michigan Water Quality Standards (WQS). The standard for PCBs is based on Rule 57 of the Michigan WQS, which requires that values be calculated to protect aquatic life, wildlife and human health, and the most restrictive of these values is used as the Michigan WQS. Thus, the standard for PCB is based on the cancer risk value to protect humans, assuming that the primary route of exposure is by consuming contaminated fish. The standard for water is 20 pg/L, which equates to a fish tissue level of 0.05 mg/kg. These values are the necessary levels to meet applicable rules and requirements for surface waters.

These remedial action goals have been coordinated with the Fisheries Management Plan (Johnson et al. 1989) and National Rivers Plan (MDNR 1981) for the Kalamazoo River. Specifically, in coordination with the Fisheries Management Plan, carp, and smallmouth bass have been identified as the target species for analyses in deter-

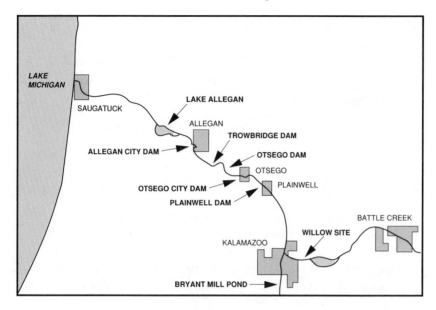

Fig. 23. Major areas of PCB sediment contamination in the Kalamazoo River basin

mining when the goals of the Kalamazoo River remedial action plan have been met.

It should also be noted that Michigan's WQS include a surface water related pathway for wildlife protection. However, for upland areas of contamination that may not impact surface waters, a terrestrial pathway may need to be developed to identify any wildlife use impairments. It is expected that the Superfund Remedial Investigation/Feasibility Study (RI/FS) will address this concern as appropriate.

In a Feasibility Study of Alternatives (NUS 1986), a PCB model was developed for the Kalamazoo River to evaluate the suitability and effectiveness of various remedial actions. All remedial investigation/feasibility studies have been undertaken as a result of a U.S. EPA 205(j) grant ($50,000) and Michigan's Environmental Response Act funds (Public Act 307: $700,000 spent to date). The Public Act 307 Program provides funds to identify hazardous waste sites, assess risks, and evaluate priorities for cleanup of environmental contamination. This program provides Michigan with the ability to take action at sites not eligible for remedies through the federal Superfund program. Recently, the Kalamazoo River has been ranked on the National Priorities List of Superfund and will be eligible to receive funding in the future. Presented below are highlights of the actions being taken at the eight areas with highest PCB contamination.

Bryant Mill Pond

Bryant Mill Pond is located on Portage Creek in the City of Kalamazoo. Bryant Mill Pond was formerly a 8.1 ha (20 acre) impoundment created by Alcott Dam. While it was impounded, it received wastes discharged from Allied Paper Company, which recycled carbonless copy paper. The river often appeared turbid and milky white from the paper company discharge.

The water level has since been lowered and the dam no longer impounds water, so contaminated sediments are exposed in the flood plain. The sediments adjacent to and in the river contain the highest concentrations of PCBs found in the Kalamazoo River basin, averaging over 100 mg/kg and ranging up to 1,000 mg/kg. A total mass of about 13,608 kg (30,000 pounds) of PCBs are contained in about 61,168 m^3 (80,000 cubic yards) of sediments at this site.

In December 1987, the State of Michigan filed suit in Federal District Court against Allied Paper Company/HM Holdings, and other property owners. The law suit seeks, among other things, remediation of this site and natural resource damages. A series of status conferences have been held to identify remedial actions, and settlement has been reached on some issues in the lawsuit. The point source discharge of PCBs has been eliminated, and consent agreements regarding contaminated landfill and dewatering lagoon areas adjacent to Bryant Mill Pond have been entered.

Several investigations have been conducted in Bryant Mill Pond under the guidance of the court. Remedial options including reimpoundment, capping and excavation have been evaluated, but a mutually agreeable final remedial action has not yet been identified. One option under review is the diversion of Portage Creek as an interim remedial action.

Willow Boulevard Site

The Willow Boulevard site is the most upstream area of PCB contamination on the Kalamazoo River. It is a former sludge disposal
area adjacent to the river owned by Georgia Pacific Corporation,
which also recycled paper. The site was used from 1965 to 1975 for
sludge disposal.

Today the site is heavily vegetated, 4.5 ha (11 acres) in size and
contains about 198,796 m³ (260,000 cubic yards) of material. There
is fairly uniform PCB contamination at about 60 mg/kg with a total
mass of about 20,455 kg (45,000 pounds) of PCBs. An important
characteristic of this site is that it has well stablized clay soils with
a permeability of 2×10^{-8} cm/sec, better than that required for
liners in PCB landfills.

Georgia Pacific has submitted a proposal to remediate the Willow
Boulevard Site (Dell Engineering, Inc. 1988). Since the primary route
of release of PCBs from this site is physical erosion, the proposed
action isolates the site and eliminates the erosion. Steel sheet piling
would be installed along the river bank and the residuals would be
covered with a multi-layered cap. In addition, Georgia Pacific would
conduct a feasibility study of onsite treatment every five years.

Plainwell, Otsego, and Trowbridge Dam Sites

The next three areas downstream are the Plainwell, Otsego, and
Trowbridge Dam sites. These areas are former impoundments
drawn down in the 1970s. The Plainwell and Otsego Dams formerly
impounded water over about 40 ha (100 acres), while the Trowbridge Dam was much larger, formerly a 227 ha (560 acre) impoundment. In 1988, the State of Michigan removed the superstructures
to decrease liability, and the intact sills of the dams maintain current water levels.

The dams were installed around 1900 and continued to impound
water through the early 1970s. PCBs were discharged to the river
primarily between the mid-1950s and the early 1970s. Therefore,
about 15 years of PCB-contaminated sediments lie over 50 years of
clean sediments. When the water level was lowered in the 1970s,
the river channel cut through the sediments which had been deposited when the area was impounded, and thus the present river chan-

nel is relatively free of contaminated sediments. PCB-contaminated sediments are now in the top 0.6 m (two feet) of exposed sediments adjacent to the river at concentrations of about 15 to 25 mg/kg. An estimated 13,608 kg (30,000 pounds) of PCBs are in the Plainwell Dam site, while 9,072 kg (20,000 pounds) are in the Otsego Dam site and 34,020 kg (75,000 pounds) remain in the Trowbridge Dam site.

The primary route of release of PCBs from these sites is erosion of sediments adjacent to the river. Michigan DNR has developed conceptual design documents for an interim remedial action to remove the contaminated sediments adjacent to the river (GZA-Donohue 1990a). The exposed contaminated sediments would be removed to create a PCB free zone of approximately 7.6 m (25 feet) on each side of the river (Figure 24). Removed sediments would be placed in spoils storage areas on adjacent Michigan DNR property. In addition, the remaining structure of the dam would be removed, which would lower the river channel and further isolate the river from the remaining contaminated sediments.

Otsego City Impoundment, Allegan City Impoundment and Lake Allegan

The Otsego City Impoundment, Allegan City Impoundment and Lake Allegan are the only impounded areas on the Kalamazoo River between the City of Kalamazoo and Lake Michigan. Otsego City Impoundment is 40 ha (100 acres) in size and has significant silt deposits. There is about 11,364 kg (25,000 pounds) of PCBs in this impoundment, with sediments at an average PCB concentration of about 25 mg/kg.

The Allegan City Impoundment is also 40 ha (100 acres) in size with PCB concentrations at about 20 to 50 mg/kg. An estimated 11,364 kg (25,000 pounds) of PCBs are also contained in these sediments.

Lake Allegan is a 607 ha (1,500 acre) lake, with PCB contaminated sediments only about 0.3 m (one foot) deep through most of the lake. The average PCB concentration in sediment is relatively low, about 10 mg/kg. However, because of the size of the lake, it contains the largest mass of PCBs of the eight areas, about 45,455 kg (100,000 pounds).

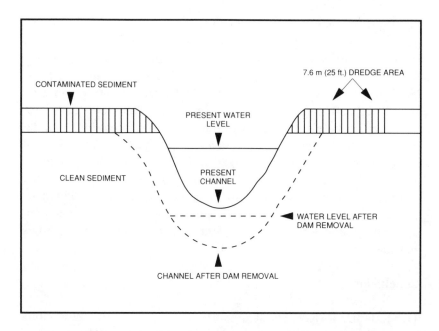

Fig. 24. Current and expected conditions after implementation of proposed interim remedial actions in the Plainwell, Otsego and Trowbridge sites

For all three impoundments, the only feasible alternative identified to date is dredging and disposal, with an associated cost of over $100 million. Uncertainties were identified in the feasibility study modeling for these three areas, including burial rates, partitioning coefficients and fish populations. Studies have been completed to address these issues (GZA-Donohue 1990b), and, based on the results of these studies and new advances in PCB remediation technologies, the feasibility study and remedial options for these three areas will be updated. It is expected that this will be part of the Superfund RI/FS.

Remedial Action Plan Progress

As noted earlier, Michigan DNR is coordinating all efforts to restore the Kalamazoo River and is working to ensure that all state, federal and international programs are complementary and reinforcing. The

State of Michigan committed to developing a remedial action plan (RAP) to restore impaired beneficial uses in the Kalamazoo River in 1985. A number of public meetings were held in 1986–1988 relative to the remedial investigation/feasibility studies and the RAP. In addition, the State of Michigan provided the City of Kalamazoo with a $10,000 grant to establish a Kalamazoo Basin Strategy Committee to provide public input to the planning process for the Kalamazoo River. The committee held a series of meetings and reported its recommendations to Michigan DNR in 1989. The Kalamazoo River RAP is in draft form (Michigan DNR 1987) and is awaiting completion of the Superfund RI/FS, expected to be completed by 1995. While the Superfund RI/FS is being conducted, continued public participation will occur as required in the State Act 307 rules and the Superfund process.

Conclusions

The Kalamazoo River is an important resource with tremendous potential for full recovery and public use in the next 10 years. However, it must be recognized that PCB contamination in the Kalamazoo River is a complex, widespread problem without easy, low cost solutions. Preliminary estimates indicate that the cost to remediate the source areas using standard technologies is well over $100 million. It is unlikely that any one source of funding will cover the entire cost of remediation.

In the 1970s, the contamination problem was still being created. The 1980s have brought the elimination of point source inputs of PCBs, quantification of the extent and level of PCB contamination, and identification and design of remedial actions. The 1990s will bring specific remedial actions to remediate PCB-contaminated sediments.

Major tools being used to address PCB-contaminated sediments and implement remedial actions include Michigan's Act 307/Environmental Bond Program, legal enforcement by the State of Michigan and the federal CERCLA process (Superfund). Keystones for future success include legal authority, additional funding and public support. Michigan DNR is dedicated to implementing all necessary cleanup actions in the 1990s to fully restore Kalamazoo River.

REFERENCES

Dell Engineering, Inc. 1988. *Report on management options at Willow Boulevard for Georgia-Pacific Corporation, Kalamazoo, Michigan.* Project No. 87678. Holland, Michigan.

GZA-Donohue. 1990a. *Interim remedial action for portions of the Kalamazoo River system—conceptual design technical memorandum.* Lansing, Michigan.

GZA-Donohue. 1990b. *Evaluating sediment burial rates and PCB partition coefficients, Kalamazoo River, Michigan.* Livonia, Michigan.

Heaton, S. 1990. *A qualitative biological survey of Kalamazoo River, near Battle Creek, Michigan (Calhoun and Kalamazoo Counties), July 7, 1988.* MI/DNR/ SWQ-90-031. Lansing, Michigan.

Johnson, D., D. Reynolds and J. Dexter. 1989. *Kalamazoo River basin fisheries management plan.* Michigan DNR, Lansing, Michigan.

Michigan Department of Natural Resources (Michigan DNR). 1981. *Lower Kalamazoo River natural river plan, Allegan County.* Lansing, Michigan.

Michigan Department of Natural Resources (Michigan DNR). 1987. *Kalamazoo remedial action plan, second draft.* Lansing, Michigan.

Morse, D. 1990. *Biological survey of the Kalamazoo River between Kalamazoo and Allegan, August 1989.* MI/DNR/SWQ-90/032. Lansing, Michigan.

NUS. 1986. *Kalamazoo River PCB project: feasibility study of alternatives.* Pittsburgh, Pennsylvania.

Chapter 13

Keystones for Success

John H. Hartig and Michael A. Zarull

> "Cleaning up the environment has become a social and political prior-
> ity. Broad-based support and participation in RAPs are essential to
> rehabilitating degraded Areas of Concern. The RAP institutional struc-
> tures which have been established for many Areas of Concern, such as
> stakeholder groups and public advisory committees, will be instrumen-
> tal in charting a future that resolves the environmental problems of the
> past and simultaneously plans for environmental and economic prosper-
> ity in the future."
>
> Gordon K. Durnil
United States Chairman
International Joint Commission

Introduction

As one can see from the preceding chapters, remedial action plans
(RAPs) are complex, not only in their attempt to address
scientifically-difficult problems but also in the novelty of their approach to achieve solutions. It took decades to manifest the degree
and extent of toxic substance contamination in Areas of Concern,
and it would be naive to think that such problems could be resolved
in a short period of time. Therefore, it must be recognized that
RAPs are a long-term process in which numerous obstacles must
be overcome to sustain progress. Continuous and vigorous oversight
is required.

Attempts under the RAP process to apply the ecosystem approach and remediate past and present damage are further complicated because restoration and economic redevelopment are occurring simultaneously in most Areas of Concern. As a result, there is
a growing need for harmonization of environmental, social, and economic development goals. RAPs provide an opportunity and a
mechanism to harmonize and achieve these interdependent goals.
At the same time, technical and community groups developed as
part of the RAP process can and must integrate their goals as rapidly

as possible with the agency or regulatory body responsible for existing remedial programs to ensure that all activities are complementary and reinforcing.

Each RAP developed thus far has experienced its own growing pains. Approximately 20 plans have been submitted to the International Joint Commission (IJC) for a Stage I review (i.e. problem identification and description of causes). While many have struggled with some of the basics, such as adequate problem definition, all that have attempted a Stage I have experienced difficulties in getting over the next hurdle—identification and implementation of remedial actions.

A successful RAP process depends on accomplishing certain key elements or cornerstones. These cornerstones form the basis upon which common problem definition is achieved, mutual goals are established, and plans to achieve these goals are developed and implemented. Even if all of these critical elements are accomplished, further major initiatives are required if the ultimate goal of restoring these Areas of Concern is to be achieved. These initiatives or requirements are the keystones for success.

Continued Public Involvement

Public involvement in RAPs has been a major breakthrough for the future of environmental protection and planning in general, as well as for ecosystem management in the Areas of Concern. Continued effective involvement can only be maintained, however, if all stakeholders are convinced that the time spent on RAPs is productive. At a RAP workshop sponsored by the IJC in 1989, participants generally agreed that if the public does not see short-term progress in remediation, the danger exists that they will become disillusioned and abandon the process (Hartig et al. 1991). Short-term, focused projects must be organized that are of interest to and achievable by the public. Building a record of such successes is one mechanism of sustaining public involvement.

Governments must convince the public that governments truly work for the public interest; on the other hand, the public is a key player in this effort and also has the responsibility to remain committed even though the going may get tough. Citizens should not accept or be expected to passively accept a dictated timetable; in the absence of suitable progress, moreover, the public should pursue creative approaches to expedite change.

It is also important to recognize that members of the public can play an important role in "holding decisionmakers' feet to the fire." Because citizens can be invaluable in keeping the RAP process focused and achieving greater accountability from all interests, continuous and vigorous public oversight of RAPs is an essential keystone to overcome bureaucratic inertia.

The development and implementation of RAPs is a long-term process, where RAPs will be updated based on additional data, and new technologies for remediation (Hartig and Thomas, 1988). Therefore, continuity in the RAP process is essential. For some RAPs, new state or provincial RAP coordinators have been appointed each year. Incredible demands are placed on these individuals, and they are regularly expected to be technically adept and publicly adroit. They frequently find themselves under severe deadlines, at the center of agency and public conflicts, and even more stressful, in most cases this is not their only job responsibility.

Because of high turnover in some RAP coordinators and numerous, difficult responsibilities within agencies, continued public involvement and a long-term commitment to RAP institutional structures (i.e. citizens' advisory committees, stakeholders' groups, basin committees, etc.) provide an important role in ensuring continuity in the RAP process (Hartig et al. 1991). No one will work harder or longer to secure their own future than the residents of each Area of Concern. The public is a unique, invaluable and free RAP resource; without continued and committed public involvement, RAPs will not succeed.

Effective Communication and Cooperation

When various regulatory, management, and public entities try to understand one another and work together to restore a given geographic area, effective communication and cooperation become essential. RAPs have been described as an experiment in institutional cooperation, where institutional structures (e.g. stakeholders' groups, basin committees, citizens' committees, public advisory committees) have been established to broadly represent municipal/local interests, elected officials, industry, land use planning agencies, fisheries/wildlife management, the public, and others. As each institutional structure attempts to reach general agreement on problems, goals and actions for their Area of Concern and to explicitly account for the interrelationships among these diverse interests in a cooperative and integrated approach to management of the resource, establishing and sustaining effective communication and cooperation become essential.

Not all citizens are knowledgeable or comfortable discussing technical aspects of the problems. Therefore, scientists must develop communication skills and be willing to help stakeholders understand technical aspects of the issues in order to reach agreement on a course of action. One option to encourage this communication link is set up a basinwide public facilitator group, that could facilitate public participation in RAPs. The IJC could assist by publishing a semi-annual RAP newsletter to increase communication between RAP groups, developing a common set of public participation guidelines, and organizing a public advisory board to provide corporate memory and share experiences between RAP groups.

The obvious benefit of effective communication and cooperation between different individuals and organizations is in building coalitions. Coalitions are essential to elevate the priority given to specific remedial actions, to obtain funding for implementation, and to develop the "strength in numbers" or substantial agreement and trust between diverse interests needed to take action. Mistrust can be common between stakeholders in the initial stages of some Areas of Concern, but this distrust can dissolve over time as the coalition develops and strengthens its sense of a united group with mutual goals. Indeed, Brown (1989) has shown that, throughout the world, success is more common when groups and stakeholders join forces to accomplish well defined, mutually beneficial tasks.

Resource Commitments

The availability of adequate human and financial resources is a common concern for all RAPs. Available resources within state, provincial and federal government environmental programs are stretched to the limit, and citizens have limited time and resources to bring to the planning process. Therefore, it is essential that all legal resources possible be provided to assist the RAP program's goals.

Some individuals have pointed out that the Great Lakes Water Quality Agreement is just that: a water quality agreement, which specifically identifies the extent of commitment between the two countries without extending into the realm of total, integrated resource management that has become identified with and essential to RAPs. Others argue that the Great Lakes Water Quality Agreement does not carry the force of law. However, the IJC and its Great Lakes Water Quality Board have recognized that RAPs are an integrated resource management program. In order to effectively identify and implement remedial actions for the complex problems we are presented with in Areas of Concern, RAPs should be incorporated into law, either by amalgamating RAPs/Areas of Concern with existing statutes or by developing new statutes (Hartig et al. 1991).

Such statutes must include the direction, authority, and funding to develop and implement RAPs. The U.S. federal Great Lakes Water Quality Improvement Act of 1990 provides direction, deadlines, and some resources for RAPs. The National Wildlife Federation has petitioned to list U.S. Areas of Concern as "toxic hotspots" under Section 304(1) of the Clean Water Act, which would make RAPs the legally enforceable control strategy required by U.S. law. These statutes and initiatives are positive steps forward in formalizing the RAP process into federal, state, provincial, and local environmental management structures.

In the short term, it is essential that existing laws be expeditiously enforced so that those polluters responsible for creating the problems will finance cleanup and remediation through fines, penalties, and consent agreements. Most often fines and penalties go into government "general fund" programs; in the future, these fines and penalties should be specifically designated for remediation in the Area of Concern. In the Ashtabula River Area of Concern, for example, the fine received from a chemical industry for a permit violation went directly to the Ashtabula RAP Advisory Council for the plan's development. In 1990, USX agreed to pay $34.1 million to help clean up the Grand Calumet/Indiana Harbor Canal Area of Concern (i.e. $26.6 million to stop discharges; $5 million to help clean up sediments; $2.5 million for studies). Such agreements should serve as models to obtain resource commitments in other Areas of Concern.

The continuing climate of federal, state and provincial government financial restraints means that limited resources are available for RAP development and implementation. In those cases where Great Lakes states and the Province of Ontario have more than one Area of Concern, RAP development has been staggered. This strategy was necessary because of the escalating costs associated with RAP development. Despite this, and because the process of development has taken rather longer than originally anticipated, most RAPs continue to be inadequately funded.

In some cases, lack of adequate government funding has been addressed by direct and novel action on the part of the local citizens. Private, nonprofit organizations are being established for some RAPs (e.g. Friends of the Rouge, Friends of the Buffalo River, Saginaw Bay Alliance) to host fundraising events and seek foundation grants. Funds raised by the organizations are used to finance the development of the RAP and its associated public involvement and communication programs. The Friends of the Rouge, for example, raises enough money to support a $100,000 annual budget to undertake public education and participation in restoring the Rouge River in southeastern Michigan. In Cleveland, Ohio, funds received from the George Gund Foundation pay for specific projects related to development of the Cuyahoga River RAP.

All of these novel, citizen-based financial schemes so far have taken place in the United States. While the cultures of the two countries are very similar, striking differences occasionally emerge, such as the activist or action-based attitude that is more prevalent in the United States. In this instance, it may be that Americans have more experience and alternatives to draw upon. Whatever the reason, it is evident that Canadian citizens involved in the RAP process will have to learn from their American neighbors. Their resulting efforts and success in fundraising will provide not only a greater independence in all facets of planning, but also the potential to support inadequately funded or slow-to-be-implemented remedial measures.

In the RAP institutional structures established in 33 of the 43 Areas of Concern thus far, committed individuals not only raise funds but also volunteer their time to attend meetings, help write sections of the RAP, liaison with other groups, organize RAP events, educate children through school programs, and work to achieve implementation. This volunteer public service is absolutely essential in developing a coordinated response to complex environmental problems and ensures that the proper societal priority is placed on solving these problems. In some areas, the atmosphere of community cooperation abounds, universities and community organizations have provided free office and meeting space, and local businesses and the industrial community have directly contributed funds to sustain the RAP process.

Every effort also must be made to apply the ecosystem approach to financing RAPs. Care must be taken to ensure that the financing of remediation is based on the concept of "polluter pays" and that attempts have been made to be equitable. A U.S. federal consent agreement for Michigan's highest ranked Superfund site (Liquid Disposal Incorporated in the Clinton River Area of Concern), for example, specified $23 million of remediation (i.e. excavation of contaminated soils, solidification, containment of solidified waste, and contaminated groundwater treatment). Over 500 industries which used the facility in the 1960s and 1970s have agreed to share the costs of this remediation, based on the amount of hazardous waste they disposed of at Liquid Disposal Incorporated.

Finally, one must remember that federal governments, in cooperation with the state and provincial governments, have the formal responsibility to provide leadership in RAP development and implementation. Governments have the mandate to enforce laws and the ability to manifest leadership through allocation of resources. Indeed, Durning (1989) has shown that only governments have sufficient resources and authority to create the conditions for full-scale grassroots mobilization required for ecosystem restoration.

Essentially, what is needed is the combination of a "carrot" (i.e. incentives) and a "stick" (i.e. disincentives) approach to RAP planning, implementation and integration. The federal governments, as parties to the Great Lakes Water Quality Agreement, must provide leadership through financial incentives and technical resources to the state, provincial and local governments to support the RAP process. In addition, penalties or sanctions must be applied for not undertaking required planning, implementation or integration. One example of such a sanction is the U.S. Environmental Protection Agency's withholding of Clean Water Act funding through Sections 106 and 319 if state governments do not meet RAP development deadlines, implement recommended remedial actions, or

ensure necessary integration between and within different programs. Such a "carrot and stick" approach is necessary to ensure
continued progress.

There is no doubt that greater federal and state/provincial resource incentives will be needed to assist local communities to do
what is, first and foremost, in their own interest (Hartig and Vallentyne, 1989). Polluters and citizens also must recognize that they
must pay their fair share of the cost.

Research Needs

Because the RAP is iterative, plans must be updated periodically to
reflect new data, information or new technologies. To be able to
truly get to the causes of complex problems such as toxic substances
contamination and ensure that remediation moves forward, an ongoing priority should be to address research needs. While a total and
comprehensive understanding of ecosystems is unlikely to be
achieved because of uncertainties caused by chance and the complexity of interactions, the point at which sufficient scientific evidence exists for a rational decision to be made must be determined
(Thomas et al. 1988). Uncertainty will always remain, but must not
be allowed to delay environmental decisions on the basis of a perceived need for more research, greater resolution of process and
understanding, or supposed reduction of the level of uncertainty.

RAP research needs can be generally categorized into four areas.

1. Ecosystem Health and Dynamics: Clearly, more research is
   needed to understand how ecosystems function and the impact of various activities on those ecosystems. The biological
   fate and effects of toxic contaminants, for example, are of primary interest in Areas of Concern. Understanding the role of
   various ecosystem compartments as sources, pathways, and
   accumulators of contaminants is vital to making decisions
   on how to adequately rehabilitate and manage the ecosystem.
2. Proactive Control: Experience has shown that the best way to
   reduce the amount of persistent toxic substances entering the
   Great Lakes is to proactively control them at their source
   (Hartig and Zarull, 1991). The United States and Canada (1987)
   explicitly recognized this when they adopted the long-term

goal of "zero discharge" of persistent toxic substances in the 1978 Great Lakes Water Quality Agreement. The Agreement recognizes that zero discharge can only be accomplished through a reduction in the generation and use of persistent toxic substances. To help implement this proactive approach, research is needed on preventing the production of new hazardous wastes and to estimate the hazards associated with those which already exist. Research should also focus on controlling contaminants at or near their source through reuse, recycling, recovery, and waste exchange.

3. *In situ* Remediation: There is no doubt, from a societal perspective, that a proactive or preventative approach to persistent toxic substances is considered the top priority. However, in numerous Areas of Concern reactive methods also will be required to resolve contaminated sediment and hazardous waste site problems in order to fully restore impaired beneficial uses. Research is needed to develop and demonstrate remediation technologies for contaminated sediments and hazardous waste sites, and techniques to protect existing critical habitats in Areas of Concern and to restore habitat lost to environmental degradation and development.

4. Socio-Economic Considerations: Much of the research and applied studies in Areas of Concern has been performed in the natural or "hard science" areas. In contrast, relatively little research has been performed in the social or "soft sciences" on resolving problems in these areas. If effective solutions to the complex environmental problems in Areas of Concern are to be found, greater emphasis must to be placed on socio-economic research and application as it relates to developing and implementing environmental solutions.

One priority should be to quantify the benefits of restoring Areas of Concern. For example, what would be the economic and social benefits of having a fishery that was safe for unlimited human consumption in the Cuyahoga River (Cleveland, Ohio) or Grand Calumet River (Gary, Indiana)? What would be the economic and social benefits of remediating contaminated sediments in the St. Clair River (Sarnia, Ontario) or the St. Louis River (Duluth, Minnesota), or having bathing beaches open continuously throughout the summer in Toronto, Ontario?

Another important area of socio-economic research is on RAP feasibility, acceptance, implementation, and funding. Given the present understanding and record of performance in cost-benefit analysis, and social values and impacts, more research is needed to identify new and creative ways to ensure that socio-economic considerations are adequately addressed. These should not hinder RAPs, but rather facilitate their implementation. Unfortunately, despite some interest, there seems to be a limited capability to complete the social/economic/institutional research necessary for RAPs in the near future.

There is no doubt that both fundamental and applied research is needed for RAPs. However, the public generally is tired and skeptical of the call for more research, and wants immediate action. Therefore, the distinction must be clear between research that seeks to obtain answers to well-defined questions (which would proceed under a hierarchical program with adequate financial support and a specific time frame) and so-called research that is used as an excuse for inaction (Zarull 1990). Governments must demonstrate this difference because public confidence and trust are at stake.

Building a Record of Success

A record of success must be built into the RAP process to keep the momentum going. For most Areas of Concern, RAP institutional structures identify short-term remedial actions to help build a record of success, and undertake long-term strategic planning to acquire the necessary data and resources to identify and implement remedial actions for more complex problems (e.g. contaminated sediments). The success of this RAP process is dependent on the ability to demonstrate progress in order to sustain public confidence and support (IJC 1989).

In the Fox River/Green Bay Area of Concern, the Green Bay Metropolitan Sewerage District (GBMSD) voluntarily reduced phosphorus and ammonia discharges as an important step toward meeting RAP goals (University of Wisconsin-Green Bay 1990). This action was celebrated by presenting GBMSD with the first "Clean Bay Backer" award. In the St. Clair River Area of Concern, Dow Canada voluntarily separated its waste streams at its Sarnia, Ontario facility (R. Denning, personal communication, Lambton Industrial Society, Sarnia, Ontario), which will allow the company to recover, reuse and recycle process wastes, thereby virtually eliminating the potential for spills and harmful discharges.

In the Rouge River Area of Concern in southeastern Michigan, stakeholders organize an annual basinwide Rouge Rescue, where over 2,000 citizens help remove debris and log jams from the river. A Rouge River Interactive Water Quality Monitoring Project involves 52 high schools in monitoring the river's water quality. Each year, students also meet for a Water Quality Congress to share monitoring results and recommendations for action with the public and key decisionmakers. In Thunder Bay, Ontario, Canadian Pacific Forest Products (CPFP), one of the largest pulp mill complexes in North America, has voluntarily committed to rebuild its mill (J. Vander Wal, personal communication, Ontario Ministry of the Environment, Thunder Bay, Ontario) to produce higher quality products at more competitive prices, and also to achieve state-of-the-art pollution controls that will support a healthy fishery in the Kaministikwia River.

These are only a few examples of success stories in Areas of Concern that help to maintain RAP momentum. Indeed, the Great Lakes Water Quality Board of the IJC believes that RAPs should identify specific milestones that can be used to measure achievement (IJC 1989). Such milestones can be specific remedial actions, pollutant loading reductions, studies completed, public participation, or biological recovery. To be most effective these indicators of progress also should be made visible to the public.

Two useful ways to publicize a record of success to the public are through annual RAP progress reports and annual "state-of-the-RAP" events (Hartig et al. 1991). For the Fox River/Green Bay RAP, an annual progress report is published and widely disseminated (University of Wisconsin-Green Bay 1990). For the Rouge River RAP, an annual legislative conference is held for all elected officials and the public (SEMCOG 1988). In both cases, RAP implementation progress is celebrated and outstanding RAP challenges highlighted. The IJC (1990) believes that such annual progress reports and "state-of-the-RAP" events are instrumental in keeping the general public aware of RAP progress, sustaining public confidence and support, and helping to ensure accountability.

The Ecosystem Approach and Sustainable Development

Can the ecosystem approach and sustainable development be achieved in Areas of Concern? As noted previously, the ecosystem approach promotes holistic thinking and coordinated actions; sustainable development tries to link society, economy, and environment through management which sustains the quality of ecosystems. Two immediate initiatives appear warranted to be able to ensure progress to these concepts: 1) governments must take steps to truly operationalize the ecosystem approach in regulatory and resource management programs; and 2) individuals must take personal responsibility to protect the ecosystems in which they live and ensure that their lifestyles and actions will sustain the quality of the Great Lakes Basin Ecosystem.

Although some progress has been made in establishing institutional structures to implement the ecosystem approach through RAP development, much needs to be done to identify new and creative ways to truly operationalize the ecosystem approach in regulatory and resource management programs. Specifically, there is a need to link ecosystem theory with environmental and resource management practice. Operationalizing the ecosystem approach is a process, and given this, additional tools such as regulations, policies, changes in institutions and consistent standards must be identified.

Further, creative and innovative ways must be used to ensure that RAP solutions address the true causes of the problems, and that the remediation taken is sustainable and achieves the desired ecosystem state. Two examples include: 1) issuance and renewal of all discharge permits within a watershed at the same time in order to account for interrelationships; and 2) ensuring that sport fish stocking rates are determined based on available forage bases and that fisheries management plans explicitly address habitat loss and restoration within critical spawning and nursery ground areas.

The influence each individual has in bringing about constructive change and making a significant contribution to restoring Great Lakes Areas of Concern should not be underestimated. RAPs have opened up the environmental decisionmaking process, and individuals not only have an opportunity, but a responsibility, to participate in the restoration of Areas of Concern. All people involved in RAPs are learning; no one individual has all the answers. Therefore, each person must recognize their unique role as a stakeholder and as a member of a larger stakeholders' group working toward a common goal of sustainable development.

It has been stated many times that to truly address some of the causes of toxic substances problems will require changes in societal values and lifestyles. Each person has an opportunity and responsibility to make those lifestyle changes in the decisions and choices made each day. For example, to address the issue of chlorinated organic compounds from bleached kraft pulp and paper mills, each citizen can help to create a demand for unbleached paper products through the choices made in the marketplace. In the absence of a major natural or social catastrophe or economic revolution, the changes in our lifestyle that bring about a truly "sustainable development" will be initiated by the actions of individuals, not governments.

RAPs are literally an experiment in grassroots ecological democracy, a once-in-a-lifetime opportunity for concerned citizens, business persons and scientists to substantially influence the management of degraded areas of the Great Lakes and alter the future of this magnificent and unique resource. It should not only be seen as an opportunity, but the responsibility of every individual to become involved, to communicate concerns and values, and to ensure a full interaction in the political, public and scientific processes in the RAP program.

Conclusions

It has been said that fundamental change is brought about through a three-step process: 1) first-hand experience with the problem; 2) acquiring an understanding of the scope and causes of the problem; and 3) making a personal commitment to bring about the changes necessary to resolve the problem. Public participation in RAPs has demonstrated that substantial numbers of people have had first-hand experiences with environmental problems in Areas of Concern. Through the development of RAPs, stakeholders have and are going through a systematic process to come to a common understanding of the scope and causes of the problem. The next step will be for stakeholders to ensure that RAPs are integrated into a broader community agenda and to secure commitments to implement remedial actions. The keystones for success include continued public involvement, continuously working to achieve effective communication and cooperation, creatively acquiring resource commitments, addressing research needs, building a record of success, operationalizing the ecosystem approach within regulatory and resource management programs, and recognizing personal responsibility.

Public participation in governmental initiatives like RAPs is an inevitable result of the democratization of modern society. Caldwell (1990) believes that this is a consequence of four trends that distinguish the modern world: 1) the spread of literacy; 2) the explosive growth of information and communication services; 3) the advancement of scientific knowledge; and 4) a belief that people are capable of shaping their futures through forethought and planning. There is no doubt that within the Great Lakes Basin Ecosystem, there is more knowledge and expertise outside government than within public bureaucracies. RAPs have provided an opportunity to take advantage of this knowledge and expertise, and have certainly benefited from it.

For governments, there are rewards and risks to direct public involvement in the RAP decisionmaking process. Rewards include: 1) enlarging the resources of knowledge, skill, and insight available to government; 2) strengthening the sense of identity between citizens and their government; and 3) providing a check against official ineptitude or abuse of authority (Caldwell 1990). The risks are: 1) unwarranted interference by nongovernmental groups with orderly processes of government; 2) pressure by special interest groups for favors or decisions that are prejudicial to the interests of society as a whole; and 3) opportunities for powerseekers to usurp the role of officials charged with responsibilities for policymaking (Caldwell 1990).

The challenge, then, is to maximize the rewards and minimize risks. When considered from this perspective, one realizes that

RAPs are both an art form and a science. RAPs attempt to overcome environmental decisionmaking gridlock by developing a coordinated societal response to resolve impaired beneficial uses in degraded areas of the Great Lakes. Science can only provide part of the information; a whole series of other factors, such as social values, economic viability and stability, and public perceptions, are also required. The challenge for all stakeholders within a RAP institutional structure is to evaluate scientific information against these other factors in reaching agreement on remedial actions.

As is noted above, it is widely recognized today that the environment and economy are closely linked. The longevity of most environmental problems also makes them intergenerational. The concept of sustainable development was developed to explicitly account for environment/economy linkages and think intergenerationally. It is defined as development that meets the needs of the present generation without compromising the ability of future generations to meet their own needs (World Commission on Environment and Development 1987).

Sustainable development is rapidly becoming the global environment/economic policy of the 1990s. Both sustainable development and the ecosystem approach in RAPs attempt to recognize the fundamental roles and interrelationships of economy, society and environment in sustaining the quality of ecosystems (Hartig and Hartig, 1990). Because restoration and economic redevelopment are occurring simultaneously in most Areas of Concern, there is a growing need for harmonization of environmental and societal-economic development goals. RAPs provide an opportunity to harmonize these goals and a mechanism to achieve interdependent ends. Therefore, RAPs provide a unique opportunity to not only implement an ecosystem approach, but practice sustainable development at the grassroots level in the Great Lakes basin. One could say that RAPs are an opportunity to act locally and to learn to think more globally.

It must not be forgotten that the RAP process is continually evolving. RAPs are intended to restore impaired beneficial uses in Great Lakes Areas of Concern, and to accomplish this there must be continuity of purpose and continuous and vigorous oversight. Despite all of its recalcitrance, slowness and frustration, the RAP process still represents the only comprehensive integrated resource management approach to restoring, maintaining and protecting the

Great Lakes Basin Ecosystem. Therefore, priority must be placed on this program and specifically on remedial actions; as a society we cannot afford to fail.

REFERENCES

Brown, L.R. 1989. *State of the World.* W.W. Norton and Co., New York.

Caldwell, L.K. 1990. Public participation in government: a double-edged phenomenon. *In: Public Participation and Remedial Action Plans: An Overview of Approaches, Activities, and Issues Arising from RAP Coordinator's Forums.* International Joint Commission, Windsor, Ontario, Canada. pp. 5–7.

Durning, A.B. 1989. Mobilizing at the grassroots. *In:* L.R. Brown (ed.), *State of the World.* W.W. Norton and Co., New York. pp. 154–173.

Hartig, J.H. and R.L. Thomas. 1988. Development of plans to restore degraded areas in the Great Lakes. *Environmental Management* 12:327–347.

Hartig, J.H. and J.R. Vallentyne. 1989. Use of an ecosystem approach to restore degraded areas of the Great Lakes. *AMBIO* 18:423–428.

Hartig, J.H. and P.D. Hartig. 1990. Remedial action plans: An opportunity to implement sustainable development at the grassroots level in the Great Lakes basin. *Alternatives* 17(3):26–31.

Hartig, J.H., L. Lovett Doust and P. Seidl. 1990. Successes and challenges in developing and implementing remedial action plans to restore degraded areas of the Great Lakes. *In:* "International and Transboundary Water Resources Issues," J.E. Fitzgibbon (ed.), *Amer. Wat. Res. Assn.* TPS-90-1:269–278.

Hartig, J.H. and M.A. Zarull. 1991. Methods of restoring degraded areas of the Great Lakes. *Rev. Env. Contam. Toxicol.* 117:127–154.

Hartig, J.H., L. Lovett Doust, M.A. Zarull, S. Leppard, L.A. New, S. Skavroneck, T. Eder, T. Coape-Arnold and G. Daniel. 1991. Overcoming obstacles in Great Lakes remedial action plans. *Int. Env. Affairs.* 3:91–107.

International Joint Commission (IJC). 1989. *Report on Great Lakes Water Quality.* Great Lakes Water Quality Board, Windsor, Ontario, Canada.

IJC. 1990. *Fifth Biennial Report on Great Lakes Water Quality.* Part II. Windsor, Ontario, Canada.

National Research Council of the United States and Royal Society of Canada. 1985. *The Great Lakes Water Quality Agreement: An Evolving Instrument for Ecosystem Management.* National Academy Press, Washington, D.C.

Southeast Michigan Council of Governments (SEMCOG). 1988. *Remedial Action Plan for the Rouge River Basin.* Vol. 1: Executive Summary. Detroit, Michigan.

Thomas, R.L., J.R. Vallentyne, K. Ogilvie, and J.D. Kingham. 1988. The ecosystems approach: A strategy for the management of renewable resources in the Great Lakes Basin. *In: Perspectives on Ecosystem Management for the Great Lakes Basin,* L.K. Caldwell (ed.). State University of New York Press, Albany, New York. pp. 31–57.

United States and Canada. 1987. *The Great Lakes Water Quality Agreement as revised by Protocol on November 1, 1987.* Windsor, Ontario, Canada.

University of Wisconsin-Green Bay. 1990. *Annual Progress Report for the Fox River/ Green Bay RAP*. Green Bay, Wisconsin.

World Commission on Environment and Development. 1987. *Our Common Future*. Oxford University Press, New York.

Zarull, M.A. 1990. *Proceedings of the Technology Transfer Symposium for the Remediation of Contaminated Sediments in the Great Lakes Basin*. International Joint Commission, Windsor, Ontario, Canada.

Glossary

Algae: simple one-celled or many-celled plants without true tissues; can be free-floating (e.g. phytoplankton), capable of carrying on photosynthesis in aquatic ecosystems.

Algal biomass: the total amount of algae (usually dry weight) in a given area or volume.

Algal blooms: high densities of algae, typically the blue-green algae, that are often seen as free floating mats; they can occur suddenly as a result of opportunistic growth spurts and can adversely affect water quality.

Anthropocentric: a term that considers humanity as the central fact of the universe.

Armoring: a protective coating of cobbles, boulders, or concrete on an exposed natural or human-made landform (e.g. slope, berm, beach, ridge) which prevents erosion.

Assessment and Remediation of Contaminated Sediments (ARCS) Program: a program initiated by the U.S. Environmental Protection Agency to: 1) assess the nature and extent of bottom sediment contamination in United States Areas of Concern; 2) evaluate and demonstrate remedial options; and 3) provide guidance for United States RAPs on how to address contaminated sediments.

Benthivore: an animal which consumes benthic organisms or benthic matter (e.g. white perch).

Benthos: the plant and animal community of the bottom of lakes, rivers, estuaries, and oceans.

Best management practices: methods, activities, maintenance procedures, or other management practices for reducing the amount of pollution entering a water body. The term originated from the rules and regulations developed pursuant to Section 208 of the United States federal Clean Water Act.

Bioaccumulate: the process of accumulation and concentration of chemical substances, which are present in the environment, in the flesh of organisms.

Bioassay: test used to evaluate the relative potency of a chemical by comparing its effect on a living organism with the effect of a standard preparation on the same type of organism.

Biochemical oxygen demand (BOD): a measure of the quantity of oxygen utilized in the biochemical oxidation of organic matter in a specified time and at a specific temperature.

Biotic communities: various species of plants and animals living within a certain habitat, each occupying a specific position in this particular environment.

Biomass: total amount of all organisms (usually dry weight) in a given area or volume.

Boundary Waters Treaty: a 1909 treaty between Canada and the United States established to cooperatively resolve problems along their common border, including water and air pollution, lake levels, and other issues of mutual concern.

Canada-Ontario Agreement (COA): an agreement between the federal government of Canada and the province of Ontario which provides the basis for cooperation in fulfilling Canada's responsibilities under the Great Lakes Water Quality Agreement.

Clean Water Act: the U.S. Federal Water Pollution Control Act, initially signed in 1972 and amended numerous times in an effort to establish national programs intended to address all types of water pollution control problems.

Combined sewer overflow (CSO): in sewerage systems which carry both sanitary sewage and storm water runoff, the portion of the flow which goes untreated to receiving streams or lakes because of sewage treatment plant overloading during storms.

Comprehensive Environmental Response, Compensation, and Liability Act (CERCLA): also known as Superfund, this legislation was initially passed in 1980 by the U.S. Congress to clean up the nation's most hazardous inactive waste sites.

Confined disposal facility (CDF): a structure for storing contaminated dredge spoils.

Control orders: enforceable orders issued in the province of Ontario to dischargers to abate pollution—these orders define tasks, compliance dates by which specific tasks must be completed, and the acceptable loads and concentrations of compounds in final effluents.

Conventional pollutants: nonpersistent, nontoxic pollutants which are routinely removed as a result of primary and secondary wastewater treatment, such as substances which consume oxygen upon decomposition, nutrients, materials which produce oily sludge deposits, and bacteria.

Dibenzofurans: a group of 135 synthetic compounds which are similar in structure to dioxins—they are toxic, persistent, and are formed during manufacturing and industrial processes, as well as during incineration.

Dioxins: a group of approximately 75 synthetic compounds which are toxic and persistent—some are manufactured commercially and some formed as unwanted byproducts of some chemical manufacturing processes (e.g. production of chlorophenols), some industrial processes (e.g. bleached kraft pulp and paper mills), and incineration (e.g. burning certain plastics).

Diversity: a term used to characterize a biological community, reflecting both relative abundance and number of species.

Ecosystem: a set of interacting elements (both living and non-living) and the processes that have some interactive, integrative, and self-regulatory capabilities.

Ecosystem approach: an integrated and holistic perspective required to protect ecosystem health and integrity; attempts to account for the interrelationships among air, water, land, and all living things, including humans.

Ecotoxicology: the study of biological effects of contaminants based on field-oriented, ecologically-interpretable, biological methods.

Eutrophication: the process of nutrient enrichment that causes high productivity and biomass in aquatic ecosystems (it can be a natural process or accelerated by human activities).

Great Lakes Protection Fund: a $100 million independent endowment funded by the U.S. Great Lakes states in proportion to their Great Lakes water use. The intent of the fund is to provide a long-term funding source for Great Lakes research that will permit continued progress toward a healthier Great Lakes ecosystem.

Habitat: the region where an animal or plant naturally grows or lives.

Heavy metals: a term previously used to describe metals with large atomic weights (e.g. mercury, lead, etc.), now used (incorrectly) to describe metals in general (e.g. iron, copper, nickel, zinc, etc.).

Hypereutrophic: the state of waters typified by excessive nutrient supply and extremely high production of organic matter—these waters are usually characterized by no or low oxygen, extremely shallow light penetration, and high turbidity.

In place pollutants: contaminants which have entered rivers and lakes from municipal and industrial discharges or nonpoint source runoff, and presently reside in association with sediments.

In situ: in its original place or position.

Integrated resource management: the application of management strategies to achieve the maximum output from the optimal use of natural resources of a specific area for the benefit of citizens and their successors.

Intraspecific competition: when individuals, within any given species of plant or animal, compete for common resources.

Leachate: contaminated aqueous liquid generated by the infiltration of rainwater or snowmelt into landfilled wastes.

Listing/delisting guidelines: a set of guidelines, based on 14 use impairments identified in the Great Lakes Water Quality Agreement, used by the IJC to recommend new Areas of Concern and make determinations on whether or not Areas of Concern have been sufficiently cleaned up.

Macroinvertebrates: invertebrate organisms which are visible to the naked eye.

Macrophytes: multicellular plant life found in aquatic ecosystems, normally visible to the naked eye.

Mass balance: an approach to evaluating the sources, transport, and fate of contaminants entering a water system, as well as their effects on water quality (in a mass balance budget, the amounts of contaminant entering the system less the quantity stored, transformed or degraded must equal the amount leaving the system).

Michigan's Act 307 Program: a program legislated under Michigan's Public Act 307 of 1982 to provide for identification, risk assessment and priority evaluation for remediation of sites of environmental contamination.

Municipal-Industrial Strategy for Abatement (MISA): a program developed by Ontario Ministry of the Environment to control and reduce the amount of toxic contaminants in all industrial and municipal effluents discharged into surface waters—it has an ultimate goal of virtual elimination of persistent toxic substances.

National Pollutant Discharge Elimination System (NPDES) Permit: a NPDES permit is issued by the federal government to discharge pollution into the waters of the United States; an individual state, such as Michigan, can take over the permitting process if it can prove it has the capacity to handle it.

Nonpoint source: a source of pollution which, instead of coming in from a pipe, enters water diffusely from land or the atmosphere (e.g. runoff from farmland or waste disposal sites).

Oligotrophic: the state of waters characterized by a small supply of nutrients and hence a small organic production, high light penetration, and low turbidity (e.g. Lake Superior).

Oxygen demanding wastes: wastes which upon decomposition use up or deplete oxygen.

Partitioning coefficient: a term used to approximate a toxic substance's relative propensity for bioconcentration or accumulation in living organisms.

Persistent toxic substances: any toxic substance with a half-life (the time required for the concentration to diminish to one-half of its original value in a lake or water body) in water of greater than eight weeks.

Phosphorus: an essential food element that can contribute to the eutrophication of aquatic ecosystems—the initial limiting nutrient in most freshwater systems.

Phytoplankton: the plant portion of the plankton community which are responsible for turning nutrients and inorganic carbon into organic material via photosynthesis.

Piscivore: an animal (e.g. lake trout or salmon) which consumes fish.

Planktivore: an animal (e.g. alewife) that consumes plankton.

Point sources: sources of pollution which can be identified from a specific geographical location, such as municipal sewer outfalls or industrial discharge pipes.

Pollution tolerant: able to endure pollution.

Potentially responsible party (PRP): any individual(s) or company(s) (such as owners, operators, transporters, or generators) potentially responsible for, or contributing to, the contamination problems at a Superfund site. Whenever possible, the U.S. Environmental Protection Agency requires PRPs, through administrative and legal actions, to clean up hazardous waste sites they have contaminated.

Polychlorinated biphenyls (PCBs): a class of chlorinated hydrocarbons that have a high degree of thermal and chemical stability, are persistent and ubiquitous in distribution, and are toxic. PCBs were widely used in electrical transformers, fire retardents, heat transfer operations and other industrial manufacturing processes.

Polynuclear aromatic hydrocarbons (PAHs): a family of organic compounds, having three or more coupled benzene (six-sided) rings; this includes a number of petroleum products and byproducts.

Pretreatment: a general term used to describe the control of toxic substances at industrial sources before discharge to municipal wastewater treatment facilities—the goals of a pretreatment program include: prevent operational interference at treatment plants, prevent pass-through of pollutants, and improve plant operations for sludge and wastewater recycling.

Primary productivity: the average rate of phytoplankton photosynthesis over a distinct period of time (e.g. day, month, year).

Primary wastewater treatment plant: a municipal wastewater treatment plant which is designed to remove floating and settleable solids from wastewater.

Radionuclide: a radioactive atom which will spontaneously decay or disintegrate by emission of particles and/or electromagnetic radiation.

Remedial Investigation/Feasibility Study (RI/FS): two distinct but related studies undertaken under Superfund. The remedial investigation (RI) examines the nature and extent of contamination problems at the site. The feasibility study (FS) evaluates different methods to remediate, or clean up, the contamination problems found during the RI.

Resource Conservation and Recovery Act (RCRA): the U.S. federal solid waste disposal act of 1976 is designed to require "cradle-to-grave" management of hazardous waste by imposing management responsibility upon generators and transporters, as well as upon owners and operators of treatment, storage and disposal facilities.

Rule 57: a rule under Michigan's Water Resources Commission Act used to develop water quality-based guidelines for toxic substances in order to protect water quality and human and ecosystem health—Rule 57 guidelines are enforceable by Michigan law.

Secondary wastewater treatment plant: a municipal wastewater treatment plant which is designed to achieve primary treatment plus bacterial degradation of organic waste.

Stakeholder: a general term used to describe any individual or organization who impacts or is impacted by a situation, resource or process.

Superfund: also known as the Comprehensive Environmental Response, Compensation, and Liability Act (CERCLA). Legislation passed in 1980 by the U.S. Congress to clean up the nation's most hazardous inactive waste sites. The legislation ensures that the sites will be cleaned up by responsible parties or the government. As part of the Superfund Amendments and Reauthorization Act of 1986 (SARA), Congress appropriated approximately $9 billion for the Superfund program.

Superfund Amendments and Reauthorization Act (SARA): passed by U.S. Congress in 1986 to update and improve the old Superfund law—authorizes the federal government to respond directly to releases, or threatened releases, of hazardous substances that may endanger public health, welfare or the environment.

Sustainable development: a term popularized by the United Nation's World Commission on Environment and Development which means development that meets the needs of the present generation without compromising the ability of future generations to meet their own needs.

Toxicity: the quality or degree of being poisonous or harmful to plant or animal life.

Toxic substance: a substance which can cause death, disease, behavioural abnormalities, cancer, genetic mutations, physiological or reproductive malfunctions, or physical deformities in any organism or its offspring, or which can become poisonous after concentration in the food chain or in combination with other substances.

Toxic Substances Control Act (TSCA): a U.S. federal act signed into law in 1976 which authorizes the U.S. Environmental Protection Agency (EPA) to obtain from industry data on the production, use, health effects, and other matters concerning chemical substances or mixtures, and allows the U.S. EPA to regulate the manufacture, processing, distribution in commerce, use and disposal of a chemical substance or mixture.

Trophic State: a relative description of a lake's biological productivity, based on availability of plant nutrients and algal biomass.

Virtual Elimination: a management goal of the Great Lakes Water Quality Agreement which seeks to eliminate inputs of persistent toxic substances through treatment, prevention, and remediation.

Waste load allocation: a traditional approach to setting limits on waste discharges (industrial or other) based on an allowable concentration for each pollutant—this approach assumes that harmful effects of pollutants are a function of pollutant concentration.

Zero discharge: the philosophy adopted in the Great Lakes Water Quality Agreement for control of inputs of persistent toxic substances. It means elimination of all inputs, whether from direct discharges into waterways or the air, indirect discharges such as agricultural and urban runoff, or inadvertent discharges from leaking landfills or contaminated sediment.

Zooplankton: the animal portion of the plankton community—these organisms are the first level consumers of the phytoplankton and bacteria present in the water column.

Index

Achievements of RAPs, 17–26, 29, 52–
54, 57, 68, 83–88, 112–116, 132–135,
152–156, 179–180, 272–274
Added costs to agriculture or industry,
7, 42, 101, 128, 142
Algal blooms, 42, 180
Area of Concern, 5, 7, 9, 11, 15, 17–28,
37–38, 49, 69, 96–97, 121, 131, 140,
145, 147, 176, 185, 212, 229, 235, 245,
247, 253, 263–268, 270–273, 275–277
Armoring, 102, 104
Assessment and Remediation of Contaminated Sediments (ARCS) Program, 102, 115

Beach closings, 7, 42, 101, 128, 142, 144,
204
Beneficial uses, 7, 13, 29, 37, 40, 43, 63,
67, 93, 95, 98–99, 110, 122, 131, 140,
161, 165, 175, 188, 204, 212, 245, 251,
277
Benthos, 166, 178
Best management practices, 154, 191
Bioaccumulate, 52, 124, 145, 147, 159,
246
Biochemical Oxygen Demand (BOD), 40,
45, 189, 192–193, 219, 221
Biotic communities, 136, 166
Bird or animal deformities or reproduc-
tive problems, 7, 42, 100, 142, 146
Boundary Waters Treaty, 9

Canada-Ontario Agreement (COA), 30,
62, 66–67, 176–177
Clean Water Act, 88, 130, 137, 153, 205,
222–223, 246, 267, 269
Combined Sewer Overflow (CSO), 68,
75, 77–79, 82–83, 85–87, 93, 96–97,
102, 105, 143, 149, 153, 187, 189–190,
197, 205, 218, 220–221, 232
Comprehensive Environmental Response, Compensation, and Liability

Act (CERCLA), 121, 127, 241, 245–
246, 253, 260
Confined Disposal Facility (CDF), 88,
124, 127–130, 136, 159, 226–227
Consent decree, 87, 219–222, 224–226,
235, 242, 244–249, 264
Contaminated sediments, 28, 31, 47–48,
52, 54, 56, 67, 69–70, 75, 85, 88–89,
96, 102, 121–122, 129, 139, 149, 159,
213, 226–227, 232, 245–246, 253,
256–258, 260, 271–272
Control orders, 28
Conventional pollutants, 153, 232

Degradation: of aesthetics, 7, 42, 101,
128, 142, 144, 204; of benthos, 7, 42,
100, 128, 142–143; of fish and wildlife
populations, 7, 42, 100, 128, 140, 142,
147, 204; of phytoplankton or
zooplankton, 7, 42, 101, 128, 142, 145
Diversity, 49, 87, 127, 143, 167
Dredging, 9, 41, 68, 88, 96–97, 121, 132–
134, 149, 194, 196, 214, 226–227, 230,
232, 244, 246, 248, 259

Ecosystem, 6–7, 9, 27–28, 40–41, 45, 57,
62, 64, 95, 99, 106, 109, 166, 168, 213,
238, 269–270, 274
Ecosystem approach, 10–11, 13, 15, 31–
32, 37, 46, 50, 59, 81, 107, 114, 137,
167–168, 177–178, 187, 209, 229, 247,
263, 269, 274, 276–278
Ecosystem management, 17, 30–31, 42,
50, 171, 264, 274–278
Empowerment, 15–17, 275–278
Eutrophication: 6, 10–12, 43, 56, 144,
147, 149, 164, 189, 199; or undesirable
algae, 7, 42, 101, 128, 142, 144

Fish: consumption advisory, 40, 75, 78,
125, 137, 140, 253; tumors or other deformities, 7, 42, 100, 128, 142–143, 204

Great Lakes Protection Fund, 55
Great Lakes Water Quality Agreement, 7, 10, 15, 65–66, 71, 94, 98, 106, 131, 162, 174, 177, 187, 211, 235, 245–248, 267, 269, 271
Great Lakes Water Quality Board, 9–15, 29, 31, 37, 45, 59, 267, 273

Habitat, 6–7, 41, 43, 56, 63, 69, 75, 87, 96, 99, 103–104, 125, 128, 137, 139, 142, 145, 147, 156, 170, 176, 204, 211, 271, 274
Heavy metals, 78, 143, 171, 192, 248
Hypereutrophic, 38, 41, 143, 197

Implementation, 9–10, 15, 17, 32, 37, 48–51, 54, 65, 81, 89–90, 95, 99, 106, 108, 111, 115, 127, 145, 155, 157–158, 176–177, 191, 200, 213, 217, 242–245, 256, 268–269, 272–273
Industrial waste, 59, 93, 165, 217, 253
In place pollutants, 54, 159. See also Contaminated sediments
Institutional structures (RAP Groups), 15, 17–27, 30, 43, 65, 80, 94, 116, 131, 147–151, 168, 204–206, 230, 248, 260, 265–266, 268, 277
Integrated resource management, 30, 46, 57, 59, 151, 154, 267–278
International Joint Commission (IJC), 7, 9, 10–11, 37, 43, 45, 59, 80, 102, 106, 161, 228–230, 252, 264, 266–267, 273–274

Jurisdiction, 13, 37, 62, 68, 79, 93–94, 130, 139, 162, 185, 230, 232–233, 242, 253

Lake Erie, 8, 10, 96–97, 122, 129, 135–136
Lake Huron, 8
Lake Michigan, 8, 38, 45, 146, 150, 214–215, 218, 227, 232, 235–237, 238, 240, 247, 249, 251, 253
Lake Ontario, 8, 45, 59–60, 62–63, 69, 93, 162–163, 165, 185, 189, 194, 196, 202, 206
Lake Superior, 8
Laws/legislation, 28, 52, 69, 87, 98, 133, 154, 158, 177, 195–196, 212, 216, 233, 241, 253, 267

Leachate, 121
Leadership (government, industry, citizen, stakeholders), 5, 28, 30, 110, 118, 158, 172, 232, 249, 269
Listing/delisting Guidelines, 46, 119
Loss of fish and wildlife habitat, 7, 42, 101, 128, 142, 145, 147, 204

Macrophytes, 41, 47, 145, 166–167, 169, 171, 175, 178–180
Mass balance, 52, 102, 156, 168, 270
Metals, 24, 78, 121, 137, 192
Michigan's Act 307 Program, 28, 84–85, 89, 252–253, 255, 260
Modeling, 33, 52, 104, 162, 168, 170–172, 255
Municipal-Industrial Strategy for Abatement (MISA), 67–68
Municipal waste, 59, 93, 165, 217

National Pollutant Discharge Elimination System (NPDES) Permit -47, 77, 87, 105, 134, 136, 158, 192, 216–217, 220–224, 226, 229, 232, 254, 267
Nonpoint source, 9–10, 28, 41, 47, 54, 56, 64, 82, 85, 88, 99, 103, 105, 143, 149, 151, 154, 197, 205–206

Oligotrophic, 38
Oxygen demanding wastes, 40, 47

Persistent toxic substances, 98, 150, 158–159, 176, 270–271
Phosphorus, 6, 10, 41, 46–47, 67, 85, 128, 144, 158, 161–162, 166–170, 189, 198–201, 219, 220, 272
Phytoplankton, 41, 101, 128, 142, 145, 166, 178
Point sources, 47, 64, 75, 99, 102–103, 105, 144, 149, 152, 216, 254, 256, 260
Pollution tolerant, 141, 143, 167
Polychlorinated Biphenyls (PCBs), 40–41, 52, 75, 78, 89, 121–122, 124, 128, 140, 159, 164, 220, 222, 235–241, 243–249, 251, 253–260
Polynuclear aromatic hydrocarbons (PAHs), 40, 69, 124, 127, 143, 248
Potentially Responsible Party (PRP), 133–134, 241

Pretreatment, 192, 208, 218–219, 231–232

Primary productivity, 41, 167

Primary wastewater treatment plant, 64, 192

Problem definition, 10, 12–13, 17, 27–28, 131, 230, 264

Public advisory committee, 17–18, 21, 24–26, 30, 45, 131, 168, 170, 172, 177, 263

Public commitment, 17, 27, 107, 172, 264–265, 268–269, 275–278

Public participation, 47, 49–50, 56, 80, 106–107, 116, 119, 151, 171, 205–206, 209, 266, 268, 273, 275–276

Public responsibility, 27, 150, 268–269, 275–278

RAPs, 3, 5, 9, 13, 15, 17, 27–31, 54, 68, 79, 93–94, 119, 130, 139, 177, 185, 212, 235, 245, 247, 253, 263–270, 272–277

Record of success, 28–29, 264, 272–273, 276

Remedial Investigation/Feasibility Study (RI/FS), 124, 241, 255, 259–260

Remedial options, 65, 69, 127, 206, 256, 259

Research, 27–28, 157, 197, 201, 207, 270–272

Resource Conservation and Recovery Act (RCRA), 137, 223

Restoration goals, 167, 173

Restrictions: on dredging activities, 7, 42, 100, 128, 142, 144; on drinking water consumption or taste and odor problems, 7, 42, 101, 128, 142, 146, 204; on fish and wildlife consumption, 7, 42, 100, 128, 140, 142, 204

Rule 57, 254

Secondary wastewater treatment plant, 40, 64, 190, 192

Socio-Economic (studies, factors, considerations), 13–17, 65, 271–272, 274–278

Stakeholder, 11, 17, 25, 28, 30, 43, 47–49, 56, 65–69, 94, 173, 180, 208, 230, 263–266, 269, 273, 275–277

Superfund, 28, 102, 121–122, 127, 131, 203, 241–242, 245–249, 252, 255–256, 259–260

Superfund Amendments and Reauthorization Act (SARA), 28, 102, 121, 127, 241–245, 253

Sustainable development, 1–2, 60, 177, 274–278

Tainting of fish and wildlife flavor, 7, 42, 100, 128, 142, 146

Toxic substance, 7, 10, 12, 15, 31–32, 41, 43, 47, 52, 55–56, 64, 96, 122, 148, 153, 156, 204, 247, 263, 270, 275

Toxic Substances Control Act (TSCA), 124, 135, 220, 246–247, 254

Toxicity, 41, 47, 84–85, 87, 240, 248

Trophic state, 46, 166

Use impairment, 12, 27, 43, 76, 94, 127, 140, 142, 146–147, 204, 255

Virtual elimination of persistent toxic substances, 66, 254

Waste load allocation, 45, 47

Wastewater treatment plant, 54, 77, 85, 122, 152, 187, 191, 205, 252–253

Zero discharge, 66, 98, 158, 271